the history of photography

an overview

the history of photography

an overview

Alma Davenport

Focal Press
Boston London

Focal Press is an imprint of Butterworth–Heinemann.

∞ Recognizing the importance of preserving what has been written, it is the policy of Butterworth–Heinemann to have the books it publishes printed on acid–free paper, and we exert our best efforts to that end.

Cover photograph:
Eadweard Muybridge
from *Animal Locomotion*
1885, gelatin silver print
Museum of Art, RISD

book design: Mark Whistler
original artwork: Boston Graphics, Inc.
typesetting: Mark Whistler/Typesetting Service Corporation
cover printing: New England Book Components
printed and bound: Edwards Brothers, Inc.

Library of Congress Cataloging in Publication Data

Davenport, Alma.
the history of photography : an overview / by Alma Davenport
p. cm.
Includes bibliographical references and index.
ISBN 0-240-80041-9 (alk. paper)
1. Photography—History. I. Title.
TR15.D25 1991
770' .9—dc20 90-46561
CIP

British Library Cataloguing in Publication Data

Davenport, Alma
The history of photography : an overview.
1. Photography, history
I. Title
770.9

ISBN 0-240-80041-9

Butterworth–Heinemann
80 Montvale Avenue
Stoneham, MA 02180

10 9 8 7 6 5 4 3 2 1
Printed in the United States of America

This book is dedicated to my mother,
Alma Wilton Davenport.

Contents

Contents

Acknowledgments

Writing a history book is neither easy nor spontaneous. However, certain individuals have certainly eased the task of compiling this text.

All historical research is based on preceding work by individuals who had the foresight to record their findings. Thus, my foremost acknowledgment must be offered to those historians who recognized, early, the importance of gathering and recording information regarding the field of photographic history. These individuals include Beaumont Newhall, Nancy Newhall, Helmut Gernsheim, and John Szarkowski. More recent contributors to the field whose works have also provided invaluable source material for this volume include Naomi Rosenblum, Marianne Fulton, and Estelle Jussim.

My gratitude for access to visual material is accorded to Laura Urbanelli, Maureen O'Brien, and the staff of the Aaron Siskind Center for Photographic Study at the Rhode Island School of Design, as well as Melody Enness and Cathy Carver (for work above and beyond the call of duty) of the Education Department at RISD. Beaumont Newhall, Diana Johnson at Brown University, Laura Straus at MAGNUM, Anne Stewart at BLACK STAR, Pat Musols at The International Museum of Photography at George Eastman House, and Shelly Jurmain at the Center for Creative Photography at the University of Arizona have made the formidable task of accessing obscure imagery into a manageable occupation.

Financial support accorded the completion of this volume has come from the SMU Foundation of Southeastern Massachusetts University and the Rhode Island State Council on the Arts. Without such support mechanisms, this book, as well as countless other historical, cultural, and artistic projects, would simply not be possible.

A project of this type relies particularly on the technical support, editors, reviewers, and manuscript critics. Barbara London got me on my way. Alan Coleman was instrumental in helping me understand the vagaries of this type of project. Jim Stone was always there with his Rolodex. Karen Speerstra and Sharon Falter at Focal Press have managed this complex project beautifully. Outside editors and reviewers Alice T. Katz, Susie Shulman, John Upton, Henry Horenstein, Pat Johnston, and Aaron Siskind have accorded valuable and significant contributions and criticisms. The book has been beautifully designed by Mark Whistler.

Lastly, I would like to thank certain members of my family for their contributions. My husband, Bruce de la Ronde, has given his unflagging support. My mother, Alma Wilton Davenport, has imparted both her editing expertise as well as an appreciation of the English language. My father, Donald Davenport, taught me, while very young, that with hard work, things get accomplished.

My deepest gratitude to you all.

I-1
The Miraculous Mirror
18th C., engraving
IMP/GEH

Photography and Its Impact on World Events—An Overview

The technique of photography, or the act of "writing with light" (photo–graphy), has evolved tremendously through the 150–plus years of its existence. Photographs have been used for innumerable and varied purposes. Camera images have recorded the battles of war, a human's first steps on the moon, and Aunt Florence's annual Labor Day picnic. Photographs have been exalted in national art galleries and affixed to refrigerators. They can take the form of infrared imagery by world–famous physicists or snapshots produced by your little brother. Photography, indeed, takes on many guises.

We have been trained in this last century to accept the photograph as a factual object. We normally assume it describes realistic events. Because we see so many of these images (one source estimates that the average American sees over 15,000 camera–generated pictures, daily), we usually accept their information passively. We view them, then disregard them. Consider, for example, newspaper or television advertisements.

What many people do not comprehend is the fact that photography has, since the 1840s, played a pivotal role in shaping world events. A great number of scientific breakthroughs could not have transpired without the use of cameras. The political structure of the world would be quite different if photography had not been invented. Indeed, the very social fabric of our lives would be altered if light–sensitive film had never been put to use.

Photography as a Scientific Tool

In 1839, photography was announced as a great scientific breakthrough. Prior to that time, all light–writing (photo–graphy) was produced by working with a fleeting, traceable image formed through the use of light and/or mirrors. The apparatus used to create these traceable images was either an immobile camera obscura (literally a darkened room; Figure 1-4), or the more portable camera lucida (Figure 1-7). Both tools were used in conjuction with tracing paper, to achieve proper artistic perspective.

The great *scientific* invention was the ability to create a lasting photographic image on either metal or paper. In both cases, the image was actually formed by light rays (as opposed to being reflected onto paper, then traced) and made stable through the use of chemistry. The inventor of the first metal plate system of photography, Louis Jacques Mandé Daguerre, thought his discovery so revolutionary that he attempted to patent it. However, in a political move, the French government decided to herald the process as an example of national scientific excellence. Government members instigated a meeting of the Academy of Sciences at the Institute of France to announce the discovery. International scientific figures as well as a goodly number of press reporters were invited to witness this great new invention.

After introducing it in their homeland, the French government decided to offer the process, without cost, to *toût le monde* (the entire world). International scientists, inventors, and artists soon initiated work incorporating Daguerre's process. Within a few years, Daguerre's original pamphlet of instructions had been translated into nine languages.

Although the initial process was costly, complex, and time–consuming, over the years it has evolved from the very cumbersome metal plate system to the streamlined systems of the 1990s. Some of the processes of today do not require the photographer to even use traditional film, much less develop negatives and prints. In some cases it is all done automatically.

As science simplified the process of photography, photography also benefited the sciences. For example, in 1940, Dr. Harold Edgerton began work with millisecond flash photography. His resulting images allowed scientists to study the effects of object movement previously only imagined (Figure 12-13).

The discovery of X–rays (beams that penetrate opaque structures) was announced in 1901 as the work of Conrad Roentgen, who received a Nobel Prize for his studies. The X–ray has since become invaluable in assisting medical diagnoses and studies in microbiology. With

I-2
Timothy O'Sullivan
Wood engraving taken from the Civil War photograph,
A Harvest of Death—
Gettysburg, July 4, 1863
Printed in *Harper's Weekly*, July 22, 1865
University of Rochester through IMP/GEH

the relatively new process of computerized color X-ray photography, medical afflictions from arthritis to tumor growth can be easily identified.

The science of astronomy has been assisted immeasurably through its alliance with photography. As early as 1877, it became possible to record a complete photographic mapping of a fixed-star firmament. As telescopes (fitted with cameras) became more sophisticated, so did the astonomer's search for information about the heavens. Today, Skylab's machinery routinely emits photographs of our solar system (and beyond) for analysis and study.

Photography as a Political Tool

However, information from the heavens is not always so scientific in its orientation. More frequently, aerial and other types of photography are being used for espionage and political purposes. As such, they have contributed to the continuing *political* evolution of international relations.

During the First World War, in 1916, cameras were used during flights over enemy territories to determine troop movements. After in-flight exposure, the film was flown to headquarters, where it was developed, and the progress of troop lines and armaments were ascertained. This information proved invaluable for strategic purposes. Some experts believe battles, and even wars, were won or lost as a result of photographically generated information.

A more benign example of photography's usefulness during time of war is the tintype. Tintypes are nonreproducable photographs on pieces of tin. Many tintypes were taken of individual soldiers during the Civil War. The subjects stood proudly before a crudely painted backdrop, their uniforms proclaiming their allegiance as Yankee or Rebel (Figure 2-5). These tintypes of soldiers were given as mementos to loved ones. Conversely, many a young lady had her likeness preserved on a tintype and presented it to her beau before he left for the battlefield.

Matthew Brady's Photographic Corp followed these soldiers into the Civil War battle zones. The resultant images were transformed into wood engravings, which were then reproduced in the popular illustrated journals of the day. In reading these papers, people far from the scene of battle got their first look at the effects of armed conflict. Some of these images and engravings were so gruesome that they stoked civilian fervor against the "other side." Thus, more than 100 years ago, photographs were used as a political tool.

John Heartfield's montages of Hitler and Mussolini, produced during the Second World War, politically lampooned these tyrants and their aggressions in Europe and Africa. While these artworks were obviously collaged manipulations, and as such were never regarded as realistic, they served as a powerful political force. Many were published in popular European periodicals, and left no doubt that a faction existed which questioned the motives and actions of a number of rising political figures in Western Europe.

More realistic, unmanipulated photographs were also routinely published in order to rouse public sentiment. During the Second World War, news-oriented photographs were featured in most periodicals of the time. These images glorified the political effort, though they often showed the cost in terms of human misery. In doing so, they underscored and intensified civilian recognition of the horror of war. More recently, during the 1960s, photographs of the Vietnam conflict were published worldwide. There is a general belief that the publication of such emotive images as that of the My Lai massacre prompted the American public to pressure the government into ending its involvement in the conflict.

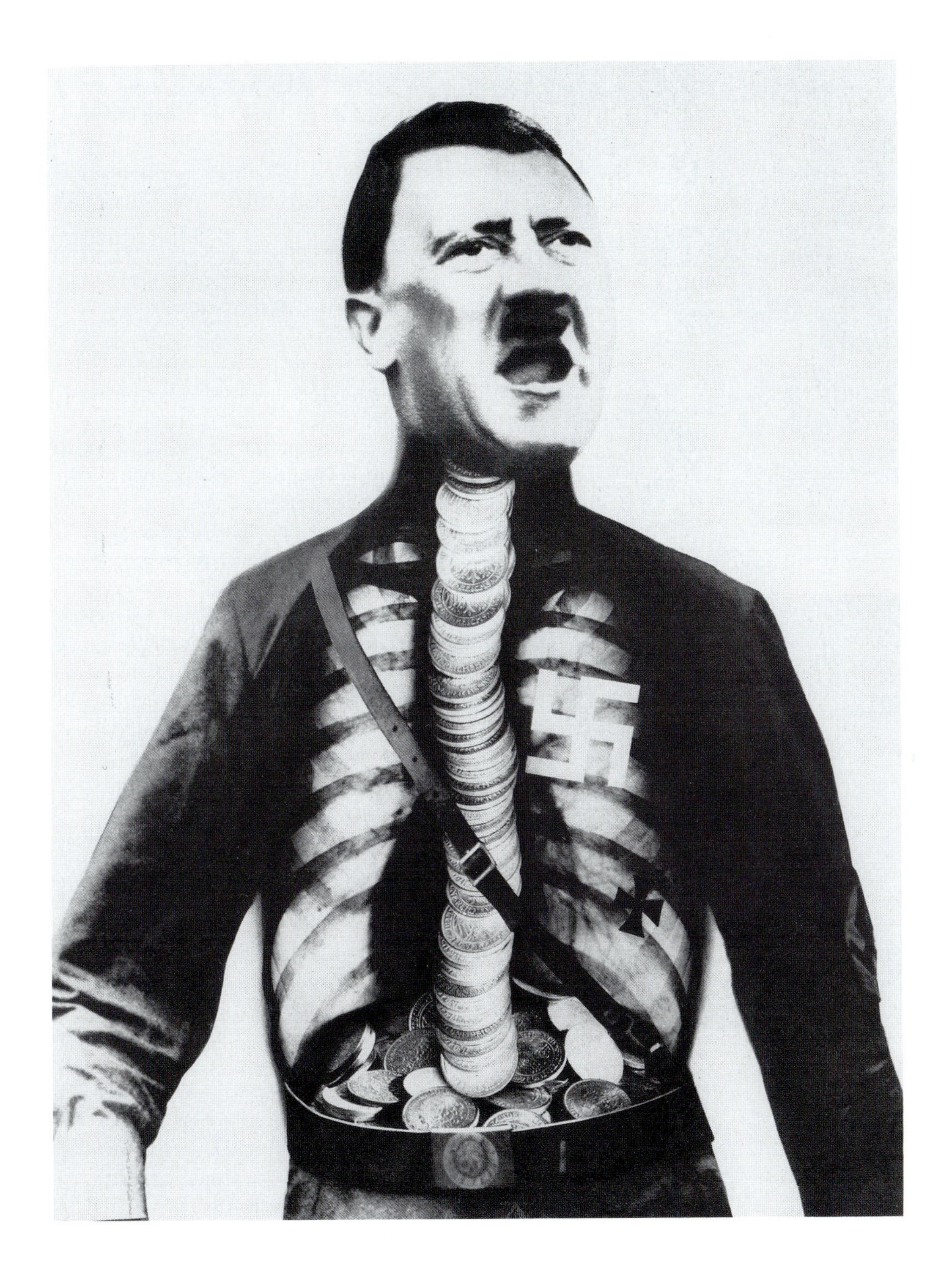

I-3
John Heartfield
Adolf, the Superman: Swallows Gold and Spouts Junk
1932, photomontage, rotogravure on newsprint
Museum of Fine Arts, Houston, Texas

I-4
Ron Haeberle
My Lai Victims
1969, gelatin silver print
©Time Warner, Inc.
Life Magazine

I-5
W. Eugene Smith
Tomoko in Her Bath
1972, gelatin silver print
©Aileen Smith and the heirs of W. Eugene Smith courtesy BLACK STAR

Photography as a Sociological and Emotional Tool

The fact that photographs provide an *emotional* impetus to alter social conditions has long been recognized. For example, engravings from photographs that documented the poor were published as early as 1851.

In that year, Englishman Henry Mayhew pioneered social awareness through photographs with his treatise, "London Labour and London Poor" (Figure 4-2). While this account did little to change the reported conditions, it started a movement in social photography that continues to this day.

Among the first sociological photographs that can actually be attributed to altering world affairs (through emotional content) are those of Lewis W. Hine. Hine started his photographic career by photographing Eastern immigrants entering the United States via Ellis Island in New York. As with Mayhew's work in London, Hine's imagery of Ellis Island did not drastically or immediately alter the conditions faced by his immigrant subjects. However, when Hine followed these immigrants into their tenement homes to photograph their living and working conditions, dramatic events eventually ensued, resulting in better conditions for his subjects, as well as for succeeding generations of the working poor (Chapter 4).

Hine's focus was eventually directed on the children of these new American immigrants. He took many photographs of young people and their abysmal working conditions (Figure 4-3 and Figure 4-4). In 1907, he was appointed as a reporter for the National Child Labor Committee. By 1916, he had accumulated enough photographic work to publish an exposé of the children and their working conditions. The report, as it centered around children, triggered an enormous emotional response. The furor it instigated prompted pressure on the United States Congress. Popular demand dictated immediate action to redress these intolerable conditions. The ultimate result was the formation of what we now know as the Child Labor Laws, which are directly designed to protect future generations of youth from such abuses.

From 1971 to 1975, W. Eugene Smith worked on a series of emotive environmental photographs when he exposed the consequences of industrial pollution in Minamata, Japan. Certainly, the emotional reaction to the physical suffering he recorded awakened his audience to the importance of enacting legislation to govern toxic waste.

Of course, not all photographs are imbued with this sense of history or emotional drama. Some imagery, notably landscapes taken during the mid–1870s by William Henry Jackson, were less emotional in their appeal. However, the power of his imagery (as well as that of lesser known photographers of the American West) prompted the United States Congress to enact legislation that created the National Parks System.

W. H. Jackson was originally employed as a photographer for a geological survey team exploring the uncharted lands west of the Mississippi River. Within a few years, they had pressed west far enough to reach, and to photograph, the Grand Canyon and the Yellowstone River. The resultant landscape images were shown to the American public (who were naturally eager to study the "Wild West"). Their reaction was so universally strong that soon members of the U.S. Congress were involved. Both funding for civilian expeditions and the creation of a national parklands system (starting with Yellowstone National Park) were a direct result of W. H. Jackson's photographic work (Figure 5-6).

Conclusion

Of course, many millions of photographs have been taken in the last 150 years. Why does history remember some and not others? Three elements distinguish photographs as visual–historical documents: composition or aesthetic value, a statement or a message, and historical or topical context. Of course, because this book focuses on the history of photography, time placement is all important. Also, because photography is a visual medium, aesthetics and composition are very important considerations. The statement or message, while assisted by composition, must also be forceful enough to communicate the idea behind the photograph.

In this text, the reader will see photographs as simple as flowers in vases. Such images cannot be evaluated on the basis of their scientific, political, or sociological merits. However, these images *do* reflect, in some way, the attitude and the prevailing notions of art and aesthetics at the time in which they were created. They are well–composed images that evoke the mood of a society; thus, it is important to include them in historical and artistic contexts.

Certainly, not all photographs *alter* world events. However, as discussed throughout this text, photographs have the ability to *shape* our scientific, political, and emotional worlds.

1-1
Joseph Nicéphore Niépce
World's first photograph—
View from Window at Gras
1826, heliograph
Gernsheim Collection, Harry Ransom
Humanities Research Center,
The University of Texas at Austin

I

From Photo–Graphy to Photography: 500 B.C. to 1840 A.D.

There are only two problems in photography.
(The first) is how to conquer light.[1]
—Edward Steichen

It was in China during the fifth century B.C. that a man named Mo Ti recorded his observation of light rays and their ability to project a "duplicate" image. He noticed that when light reflected off an object, and that reflection passed through a pinhole onto a dark surface, an inverted image of the object was evident on the darker surface. The light was "writing" a description of the object. Thus begins the story of photo–graphy, which translates from Greek to "writing" (graphos) with "light" (phot–).

Because we do not know the circumstances that surrounded Mo Ti's discovery, we can only guess that the image he saw projected was bright, but inexact—more like a fuzzy sketch than a perfect rendering—because the diameter of the pinhole that the light passed through was probably rather large (Figure 1-2). It was not until the tenth century that an Arabian, Ibn Al–Haitham (Alhazen), repeated this visual experiment and realized that by reducing the diameter of the pinhole, a fainter, though more precise, reflected image resulted (Figure 1-3).

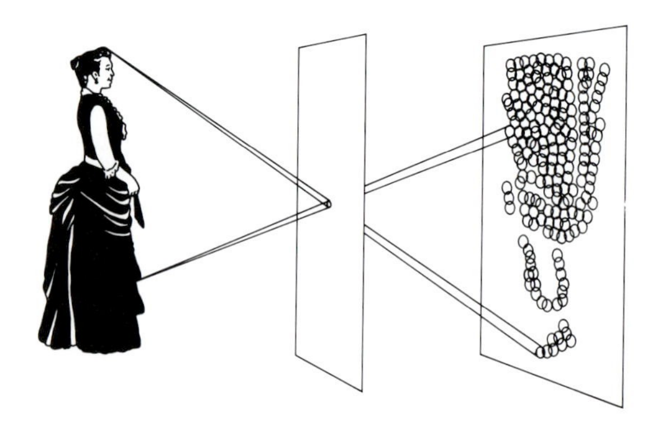

1-2
An example of the first photo–graphic discovery. When light reflects off an object and passes through a pinhole, it creates an image of the original object. When the pinhole is large, the reflected light rays have the opportunity to spread, and are received on a flat surface as large circles. These "circles of confusion" run into each other, and create an unclear image.

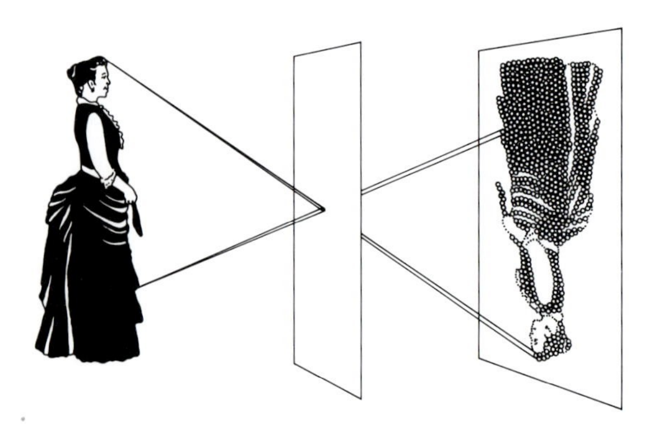

1-3
When the size of the pinhole is reduced, the amount of light reaching the flat surface is reduced and, therefore, spreads less than if the pinhole were larger. The resulting image is visually soft, but readable.

The Camera Obscura

Centuries later, this principle of photography (light writing) was further refined through the use of the *camera obscura* (dark room). These darkened rooms were used by people who wanted to view and record exterior scenes from an interior vantage point. By using this system, studies could be made of the heavens, the passage of the seasons could be recorded, and architectural studies could be made. In the 1560s, an Italian, Danielo Barbaro, noted that a lens could be substituted for a pinhole, resulting in a further sharpening and brightening of the image. Revisions of the original camera obscura concept continued into the late seventeenth century, when interested individuals constructed cumbersome tentlike rooms in order to increase the camera obscura's portability.

By the seventeenth century, the camera obscura had become a familiar tool for artists, draftsmen, and scientists. Philosophers were describing the images produced through use of the camera obscura as a perfect duplicate of life itself. In retrospect, the proclaimed perfection of images rendered through use of the camera obscura was somewhat premature. For example, although artists used it for renderings, the perspective angles that resulted were not accurate because of the simplicity of the lens. Further, the camera obscura was rather inconvenient to use. It required an interior area large enough for an adult to stand erect without obscuring the image cast on the far wall. Because of this space requirement, even the tent types were normally too cumbersome for spontaneous use.

The two problems of portability and image accuracy were resolved in the 1700s. First, the problem of mobility was overcome by minimizing

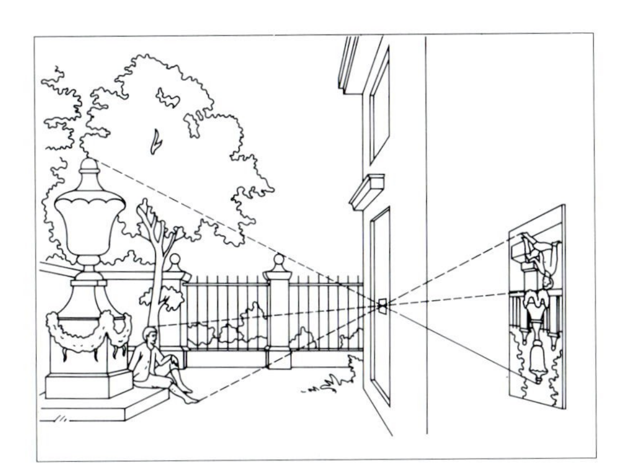

1-4
An example of a 15th century immobile camera obscura. By the 1700s, refinements allowed artists to carry this photo–graphic device into the field to obtain clear reflected images for artistic rendering.

the required space. The camera obscura was reduced from a fixed room to a small closet, and finally to a 24–inch rectangular box (Figure 1-5) that could be easily carried into the field to produce any type of artistic or scientific rendering. This box was fitted with a simple magnifying–type lens at one end and a sheet of frosted glass at the other end. The view to be reproduced would pass through the lens and be reflected on the ground glass. The artist or scientist would place a piece of translucent paper on the ground glass, and simply trace the reflected image.

The second problem, that of representational accuracy, was also resolved in the 1700s. Prior to that time, the image cast by the lens was not an accurate depiction of the scene. The simple, single–element lenses that were used produced convex, dish–shaped images. When these were received on the flat surface of the ground glass they reflected distortions of the subject. As the painters and scientists of that time insisted on perfect perspective representations, it was obvious that a more sophisticated lens system was required. Although there is no exact date for the invention of a flat field lens, or one that minimized distortions, it must have occurred between 1735 and 1768, the years in which Giovanni Canaletto produced most of his paintings.

It is well documented that Canaletto made extensive use of the camera obscura. His early work evidences distortion from the use of a single–element lens. His later work has been described as "the sacrificing of pictoral freedom to greater precision."[2] It is probable that this enhanced reality was due to the introduction of a more sophisticated lens system in his camera obscura.

Alternatives to the Camera Obscura

Canaletto was a trained artist, whereas most of the populace who wished to fabricate images through the use of light were amateurs. As the camera obscura necessitated a bit of technical expertise, alternate methods were explored so that photo–graphy could be used with little or no training required. Toward that end, silhouette machines were developed and were in use by the 1780s (Figure 1-6). An object or sitter would be positioned between a bright candle (or other light source) and the silhouette–maker. The object's shadow would fall on a piece of paper held in an easel. The silhouette maker would then cut around the edges of the cast shadow, which resulted in an image that was usually quite true to life.

Silhouettes were primarily used for portraiture, as was another system for rendering human likenesses—the physionotrace. Again, the sitter's profile was "traced," although this method used copper plates rather than paper. These plates were then engraved, which allowed for duplication of the portrait—a feature that was unavailable through the cut–paper method (Figure 1-10).

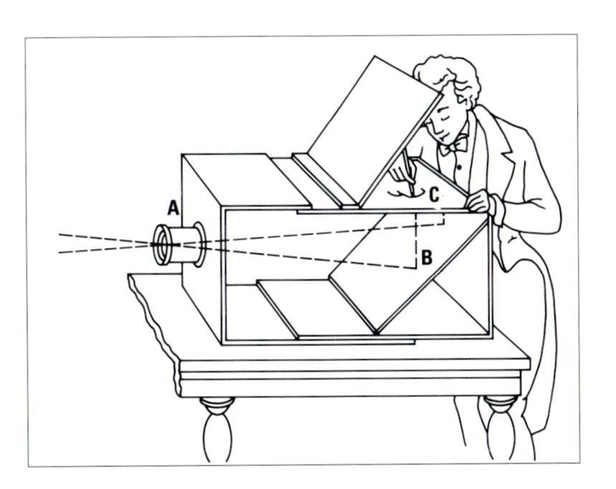

1-5
A portable camera obscura. Light reflecting off an object is gathered by the lens (A), reflected by the mirror (B), and cast onto the ground glass (C).

1-6
A silhouette–making apparatus.

Another refinement of the original camera obscura concept occurred in 1807, when William Hyde Wollaston constructed the *camera lucida.* Wollaston named his invention well, as *lucida* is derived from the Latin word *lucere,* which means to shine (light). The camera lucida was a simple, portable apparatus (Figure 1-7). Light reflected off a subject and shone through a prism anchored to a metal rod. This rod was attached to a board on which drawing paper was laid. Through the use of a peephole adjacent to the prism, the subject and the paper could be viewed simultaneously. The "artist" simply transcribed on paper the view seen through the prism.

Although the camera lucida made renderings a bit easier than the camera obscura, its chief advantage was portability. During the first half of the 1800s, an ever–increasing international curiosity was developing. Privately and publicly funded expeditions, led by artists and scientists, ventured to the Near East and beyond. Camera lucidas and small camera obscuras became as indispensable as notebooks, allowing voyagers to complete visual diaries as well as written ones.

Use of tools such as the camera obscura, camera lucida, the physionotrace, and silhouettes should certainly be considered photo–graphy, as reflected *light* was used to *write* (or draw). However, up to this time, a person was the instrument who physically created and finished the rendering. *Light* was not creating the permanent record. The age of photography (as opposed to photo–graphy) had not yet begun, for there was no method or substance yet discovered that could collect light and etch the reflection permanently onto an object.

Arresting a Purely Photographic Image

Although experiments with chemical substances such as silver nitrate, which reacted to light rather than heat, were performed throughout most of the eighteenth century, it was not until just prior to 1800 that man was first able to use light to describe an object. Thomas Wedgwood, son of the famous British potter, was commissioned by Catherine the Great of Russia to provide a china table service decorated with more than 1,000 views of country homes and gardens. First, the sites were meticulously drawn using the camera obscura. Then, since Wedgwood was familiar with the recent experiments using silver salts, he attempted to transcribe the sketches onto various surfaces using nitrates and light. He did, in fact, produce an image that was a direct result of the sun's rays. He describes the surface in this way: "...on being exposed to day light, it speedily changes color, and, after passing through different shades of greys and brown, becomes at length nearly black."[3] Unfortunately, these "sun prints" were not permanent, as the unexposed silver salts were unstable. Further illumination required for viewing the print, unless very dim, quickly exposed the unstable silver salts and permanently obliterated the entire image.

Joseph Nicéphore Niépce

In 1827, Joseph Nicéphore Niépce, a Frenchman, was the first to stabilize the camera's image. He did this by coating a metal plate with bitumen of Judea, then placing the coated plate in a camera obscura. He pointed the camera out a window and waited several hours for the exposure to materialize. After removing the plate from the camera, he immersed it in an oil solvent. The oil removed the bitumen that had not been struck by light. It left a lasting image.

Niépce had, thus, successfully exposed, developed, and fixed an image that was directly formed by light. He called his discovery "heliography," after "helios," which is Greek for sun, and "graphos," for writing.

The discoveries of Niépce were tremendously important to the evolution of photography, but they had two fatal flaws. The first was the low sensitivity of the bitumen emulsion he was using. Exposures in a camera obscura could take as long as three days. His most familiar image (and one which is generally regarded as the "first photograph") took eight hours to expose in bright daylight (Figure 1-1). Because the sun naturally shifted positions during the exposure, all shadow delineation was lost and the image was difficult to comprehend visually. The second problem was his method of applying the bitumen emulsion. Niépce used a dabber, similar to a large cotton swab, which resulted in an uneven coating of emulsion. Because the emulsion was not applied smoothly, the resultant image was blotchy in appearance.

Niépce realized he was involved with a great discovery. However, even though he retired from all other work (which was not a particular hardship as he was rather wealthy) and hired his son, Isidore, to work with him, he could not overcome the difficulties of long exposures and uneven results. He was beginning to despair and to think that his invention could be used only for minimal engraving purposes, when he received a letter of inquiry from a fellow Frenchman, Louis Jacques Mandé Daguerre.

Niépce and Daguerre—A Collaboration

Daguerre had heard of Niépce's experiments from a mutual acquaintance, Chevalier—a Parisian lensmaker. The lensmaker had advised Daguerre that a fellow countryman had ordered camera obscuras for experiments very similar to those Daguerre was attempting. Natural curiosity prompted Daguerre to contact Niépce and a correspondence was initiated. Within a year, Niépce visited Daguerre in Paris, where they discussed their mutual interests. Two years later, in 1829, they signed a ten–year partnership agreement.

Only four years of their agreement had elapsed when Niépce died. Although Daguerre urged Isidore to continue experiments in his father's name, Isidore refused, suspecting that Daguerre would ultimately steal the invention.

1-7
The portable camera lucida. This photographic device used a prism to cast a visually accurate image onto tracing paper. The convenience and portability of a camera lucida allowed artists to record foreign subject matter throughout their extensive travels.

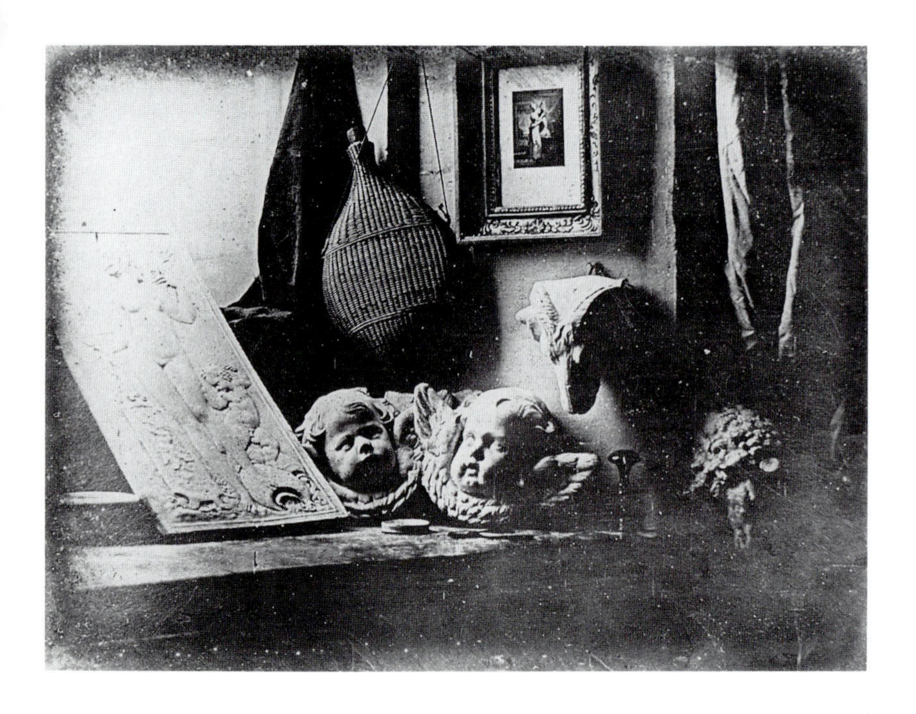

1-8
Louis Jacques Mandé Daguerre
Still Life
1837, daguerreotype
Société Française de Photographie

It is debatable whether Isidore was correct in his suspicions of Daguerre. It is known that after the death of Niépce, Daguerre offered Isidore a new contract that stated clearly that he (Daguerre) was to be considered the sole inventor and would only transfer the partnership if "this new process shall bear the name of Daguerre alone."[4] Further, it was stated that Niépce's process could be published simultaneously, but not with the same content. Also, when the French government offered an outright award for the rights to the process, Daguerre received 6,000 francs, while Isidore was awarded only 4,000.

Isidore signed the contract reluctantly and completely discontinued working on his father's portion of the invention. Meanwhile, Daguerre was proceeding feverishly in Paris, preparing a pamphlet describing the process. He envisioned his process would be sold by subscription to interested, primarily wealthy, individuals. The French government caught wind of this private enterprise and persuaded Daguerre to accept the award of 6,000 francs in order that the government could offer the process, free, to the entire world. That is, free to all the world save England. Daguerre, knowing international law, scurried hastily to England, where he received patents and, eventually, royalties on his discoveries.

The Daguerreotype—A Startling Invention

It was at the Academy of Sciences in 1839 that Daguerre gave his first demonstration of his process. The workings were complex. They required a copper sheet to be plated with silver and then exposed to iodine vapors, which produced the light-sensitive emulsion, silver iodide. The plate was exposed in a portable camera obscura, then "developed" by exposure to mercury vapors, and finally "fixed" in a bath of hyposulfite of soda. It is true that his process varied considerably from that of Niépce in the chemistry used, the shorter exposure time necessary, and the resultant crispness of the image.

After the demonstration, Daguerre's name was immediately on the lips of people around the world. Reports issued from the halls of the Academy of Science immediately following the demonstration. The reactions were universal: "great discovery!;" "extraordinary result!;" "this is Nature itself!"

William Henry Fox Talbot—The Calotype

The news of Daguerre's invention startled the world and astonished perhaps no one more than British scholar William Henry Fox Talbot. Unaware of Daguerre's discoveries, Talbot had also been working on a process that would permanently arrest the image of light on paper. In 1833, (six years before Daguerre's demonstration), Talbot had conceived a method of making permanent photographic images. Within 2 years, he had devised a light–sensitive emulsion by making alternate coatings of sodium chloride and silver nitrate on paper. These chemicals reacted with each other to form silver chloride, a light–sensitive material. Talbot placed simple natural objects in contact with this sensitized printing paper, then exposed it to the sun. Where the paper was not protected by the opacity of the object, the emulsion would darken, rendering a white silhouette against the dark background.

The accuracy of his results were startling. As Talbot wrote, "Upon one occasion, having made an image of a piece of lace of an elaborate pattern, I showed it to some persons at the distance of a few feet when the reply was, 'That they were not to be so easily deceived, for that it was evidently no picture, but the piece of lace itself."[5] Today, these images would be called contact prints, as the negative or object had been placed in direct contact with the paper's emulsion. Talbot, however, referred to them as "photogenic drawings." They were somewhat different than a camera obscura's image, as the latter was formed through reflected light rays, not those that had been blocked by an object resting on a light–sensitive surface.

Talbot's photogenic drawings were a derivation of Thomas Wedgwood's earlier experiments conducted while attempting to decorate the china service for Catherine the Great. Talbot knew of Wedgwood's work with silver salts and he felt as though he had significantly improved on Wedgwood's theories. Talbot had, by 1835, improved the photographic paper's sensitivity (through a faster emulsion) and had succeeded in "fixing" the image through the use of either potassium iodide or salt.

Talbot continued to improve his basic process through the mid–1830s. As his first trials rendered only the highlighted portions of the objects he wished to record, he decided that the problem lay in the size of the camera obscura and the lens he was using. The camera was too large, thus diminishing the amount of light that reached the sensitized paper at the end opposite the lens. To correct this problem, Talbot began to collect small boxes and convert them into cameras—"little mouse traps," his wife called them—to photograph his home, Lacock Abbey. His first successful image, in 1835, produced a one–inch square paper negative of a latticed window, which he claimed, "might be supposed to be the work of some Lilliputian artist."[6]

Negatives and Positives—Calotypes Versus Daguerreotypes

Talbot had been working on his photogenic drawing process for more than five years when he heard of Daguerre's presentation to the French Academy of Sciences on January 7, 1839. Prior to this news, Talbot had been working on a variety of projects, not all of them photographic. He had partially succeeded in expanding from the "contact print" photogenic drawing process to using lens and paper in his "little mousetraps." However, when Daguerre's proclamation was publicized, he realized he had no time to lose. He rushed samples of his drawings to the British Royal Society, after which he corresponded with various academicians, stating that he would claim priority in the method of "fixing the images of the camera obscura and the subsequent preservation of the image so they would bear full sunlight (Figure 1-9)."[7] On January 31, scarcely three weeks after Daguerre had made his presentation in France, Talbot's paper, "Some Account of the Art of Photogenic Drawing—The Process by Which Natural Objects May Be Made to Delineate Themselves Without the Aid of the Artist's Pencil" was presented to the the British Royal Society.

1-9
William Henry Fox Talbot
Hydrangia
1840, photogenic drawing negative on a salt paper print
Lacock Abbey Collection,
Fox Talbot Museum

1-10
Gilles–Louis Chretien
Portrait of Mme. Roland
ca. 1793, physionotrace engraving
Museum of Art, RISD

1-11
Hippolyte Bayard
Self Portrait as a Drowned Man
1840, direct paper positive
Société Française de Photographie

It is interesting to note that there existed such a strong competetive feeling between the two men, especially as the processes they had developed individually were quite dissimilar. The photogenic drawing process was a "contact printing" method of making light–generated imagery. Even after Talbot succeeded in making images with a camera (which he originally dubbed Talbotypes, later called calotypes), his methodology was very different from that of Daguerre. The light–sensitive chemistry involved was different. Daguerreotypes produced images on metal. Calotypes were produced on paper. Daguerreotypes were very fragile and had to be encased in glass for the image to survive repeated handlings. Calotypes were relatively durable. Perhaps the most important difference was that daguerreotypes produced unique positive images, while calotypes used a negative system capable of producing endless duplicates. However, as different as their processes appeared, it was the *idea* of permanent, light–generated imagery that seemed to be the crux of the competition.

Hippolyte Bayard—A Losing Latecomer

Other inventors were working at this type of light–sensitive system independently of both Daguerre and Talbot. Because of the circumstances (and politics) involved, they received only a hint of recognition. Hippolyte Bayard, a Frenchman of modest means, had been working with light and emulsions since 1837. His first results were negatives produced on silver chloride paper. Like Talbot, he was stunned to hear of Daguerre's process and within months of Daguerre's pronouncement, had succeeded in producing direct positive imagery on paper. Upon showing these results to Daguerre's "mentor" in the French government, François Arago, he was given a token payment to continue experimentation. In retrospect, the meager sum Bayard was awarded smacked of "hush money," as Arago counseled Bayard to remain quiet about his discoveries.

As could be expected, when the public heard of Daguerre's invention, it superceded all others, leaving Bayard as a shadow—or as he would have it, a half–drowned corpse. Bitter at the treatment he had received, he used his process to make a self–portrait (Figure 1-11). On the back of the print he wrote, "The body you see here is that of Monsieur Bayard...The Academy, the King, and all who have seen his pictures admired them, just as you do. Admiration brought him prestige, but not a sou. The government, which gave M. Daguerre so much, said it could do nothing for M. Bayard at all, and the wretch drowned himself."[8]

Sir John Herschel's Early Contributions

Sir John F.W. Herschel, noted astronomer and scientist, was also working with the camera obscura and light–sensitive materials. Although he received satisfactory results, his most important discovery was the method of "arresting the further action" of light—or "fixing" the image. Herschel had noted, in 1819, that hyposulfite of soda dissolved unstable silver salts. When he used this discovery in a photographic context, he found that he could use hyposulfite of soda to retain images he had made. Talbot learned of Herschel's discovery and, with his permission, described it in a letter published to the French Academy of Sciences. Daguerre, being French, also heard of the process, and was quick to adopt the use of "hypo" as well. To this day, the retention of almost all photographic images relies on Herschel's discovery.

Sir John Herschel also left another very important contribution to the history of the medium. He was an amateur linguist and noted that the verbiage used by Daguerre and Talbot was somewhat awkward. For example, Talbot used "reversed copy" and "re–reversed copy" to describe his results. Herschel, hearing this, suggested substituting the words "negative" and "positive," respectively. Herschel also suggested using the word PHOTOGRAPHY to describe the entire light–writing process. That terminology seemed to best describe the results of all experimenters who were beginning to depict the world through the action of the sun on light–sensitive materials.

Innumerable improvements have been made to the original photographic process in the past 150–plus years. Exposure times have been shortened dramatically. The application of emulsions has been standardized. The machinery has been streamlined. However sophisticated the process has become, its roots lie in the contributions of many light–writing pioneers—especially Louis Jacques Mandé Daguerre and William Henry Fox Talbot.

We are...wondering over the photograph as a charming novelty; but before another generation has passed away, it will be recognized that a new epoch in the history of human progress dates from the time when He who—never but in uncreated light Dwelt from eternity—took a pencil of fire from the hand of the "angel standing in the sun," and placed it in the hands of a mortal.[9]

—Oliver Wendell Holmes

2-1
Frederick Fargo Church
George Eastman with a Kodak Camera on Board the "S.S. Gallia"
1890, albumen print
from a Kodak film negative
IMP/GEH

2

Photography: Inventions, Problems, and Solutions—1840s to 1990s

Of all the arts the one that seems most miraculous is photography. That the rays of the sun...should...be directed on a bit of paper...would certainly have been deemed impossible...yet there are large numbers of persons who by the daily performance of this miracle obtain bread and meat for themselves and little shoes and bibs for their children.

—*Scientific American*
August 9, 1862

2-2
William Henry Fox Talbot
Plate VI—Pencil of Nature
1844, salt print from a calotype negative
courtesy Hans P. Kraus, Jr., New York
Museum of Art, RISD

The year was 1840, and the world was enthralled by the photographic presentations of Talbot and Daguerre. In general, people were amazed at the process by which pictures of perfection could be formed directly through the use of natural forces. Although the public was awed by this new invention, opinions differed about how best to utilize photography. Painters employed photographs as an artistic tool—in the same manner as they handled their sketchbooks. Scientists commented that the photographic process could certainly be a great boon, but they were not quite certain in what context. Some of these concerns regarding application and use of the new process were quite understandable, as both daguerreotypes and calotypes had basic problems.

Fuzzy and Fragile—Early Technical Problems

Talbot's first invention, "photogenic drawing," had initially proven useful only for portraying relatively two-dimensional objects, for unless the object was flat, forming good contact with the printing paper, the resultant image would be blurred. Later, photogenic drawing was used for both negatives from cameras and "contact" photograms.

Talbot's calotype process was different from his earlier photogenic drawings as the photographed image needed to be chemically developed before it could be viewed. However, even after exacting development, the calotype was plagued with the problem of fuzziness. Although Talbot lessened the size of the negative through use of a smaller camera, the problem of fuzziness remained. The blurred image was caused by the use of paper as the foundation for his emulsion. As the original "photograph" from the camera was a negative, it was necessary to lay that fibrous paper onto another sheet of sensitized paper in order to receive a positive image. Transmission of light through two thicknesses of wood fibers resulted in an indistinct representation of the subject. Although the process did allow endless reproductions from the original negative, the texture produced was too visually "soft" to suit

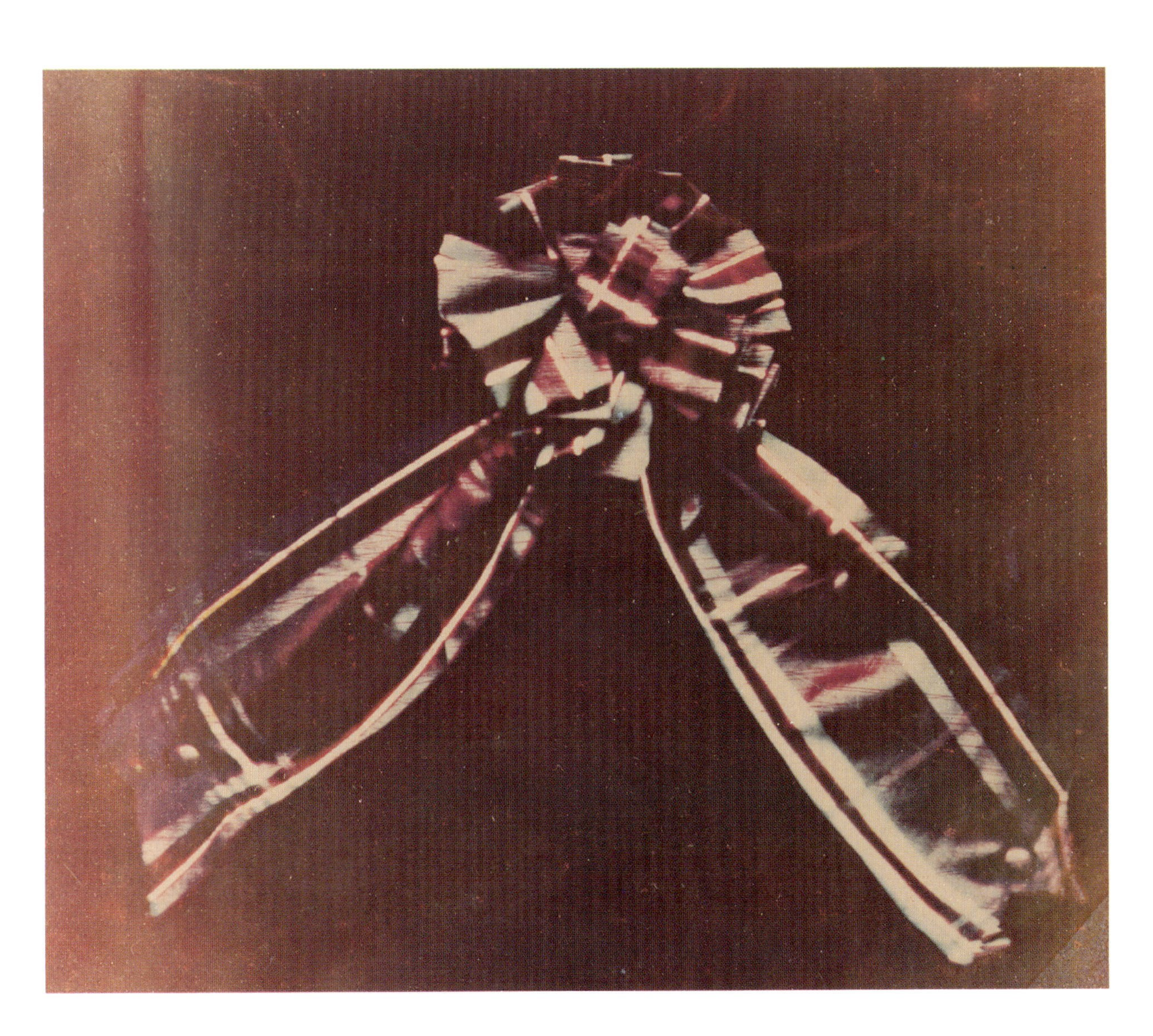

Color Plate 1
James Clerk Maxwell
Tartan Ribbon
1861, reproduction print from a photographic projection
Trustees of the Science Museum, London

Color Plate 2
Edward Steichen
Floral Bouquet in Greenhouse
ca. 1910, autochrome
reprinted with the permission of
Joanna T. Steichen
IMP/GEH

Color Plate 3
Howard McKechnie
Family Portrait
1960, Kodacolor print
private collection

Color Plate 4
Jim Stone
Breckenridge, Colorado
1978, SX–70 print

collection of the artist

most of the public. One photographer described the calotype print as the "imperfect work of man...and not the perfect work of God."[1]

Daguerre's process, while certainly producing images that were sharper and "closer to life" than Talbot's, had completely different problems. For example, daguerreotypes were one–of–a–kind objects; they could not be duplicated or reproduced. They also "reversed" whatever had been photographed. For example, any writing would be backward. Also, the emulsion coating on the plate was extremely sensitive to handling and was aptly described as "fragile as a butterfly's wing." Because of this delicacy, daguerreotypes had to be encased in bulky glass covers and leather cases for protection. Further, the process was inordinately expensive for the average person; the camera and processing equipment cost the better part of a month's salary. Last, the exposure times were intolerably long. For indoor work, in 1839, a normal exposure lasted 2400 seconds—or 40 minutes.

The inability to duplicate daguerreotypes and the delicacy of the emulsion were not major concerns for the general public. People liked the idea of a singular, thus more personal object. Hippolyte Fizeau's process of coating, or "gilding" the completed daguerreotype with gold chloride somewhat alleviated the problem of fragile emulsion. The expense of purchasing a license for the process was seldom a problem, as most people reckoned that others would purchase the licenses and would hire out their services. Therefore, the problem of lengthy exposure time was the most troublesome in the early years of daguerreotypy.

Long exposure times were particularly nettlesome in terms of certain subject matter. While some people were content to either marvel at daguerreotypes on display or to make hour–long exposures of still lifes in the studio, the overwhelming popular desire was for daguerreotype portraits.

Portrait Mania Versus Long Exposures

Prior to the 1830s, only the very wealthy could afford to have their likenesses preserved for posterity—through paintings or renderings. A person would have to expend both a financial outlay and the time required to sit for an artist. With the advent of the daguerreotype, the fledgling middle class could finally preserve their individual and personal pasts at a price they could afford, provided they could sit still long enough to have their likenesses recorded. That was not necessarily a simple requirement. For the first year of daguerreotypy, exposures lasted far too long for an average person to sit unblinking and immobile.

Many systems were attempted in order to shorten exposure times. Inventors mixed new emulsions, which they hoped would accelerate the process. Camera owners struggled to increase the level of illumination by building special rooms that incorporated mirrors and bottles full of blue liquid. These containers of liquid were supposed to shorten

2-3
unknown photographer
ca. 1845, daguerreotype
IMP/GEH

exposure times by intensifying certain portions of the sun's rays. Sitting for a portrait in such a room was an ordeal akin to torture. As one subject recollects; "(he sat) for eight minutes with the strong sunlight shining on his face and tears trickling down his cheeks while...the operator promenaded the room with watch in hand, calling out the time every five seconds, till the fountains of his eyes were dry."[2]

By 1840, a year after Daguerre's initial demonstration, Peter Voigtländer, a German, marketed a greatly improved lens. The design of this lens allowed much more light to reach the sensitized plate in a given period, thus shortening indoor exposure time from 2400 seconds to approximately 150 seconds. In today's terms, the aperture would be equivalent to an f/3.6. Voigtländer's lens attained immediate popularity. Countless imitations were fabricated in both France (where the copies were marked as "German lenses") and in America, where unscrupulous manufacturers engraved an uncapitalized "voigtländer" on each product.

A further refinement that shortened exposure time was the recoating of the plates with additional light-sensitive material, such as chlorine or bromine. Scientists referred to it as "accelerator;" others called it "quickstuff." Once this more sensitive chemical emulsion was in common use, photographic studios opened in nearly every part of the world. France, the country of origin, boasted many galleries and operators. Americans took to the new process with particular zeal. Portrait studios abounded and gallery spaces opened for public viewing of architectural, natural, and travel scenes.

Experiments with Emulsions

While most people preferred the visual clarity of the daguerreotype over the blurrier calotype, it was still essentially nonreproducible (although certain daguerreotypists did copy their plates). A combination of the two processes appeared to be the answer. The merge of the two required a resulting negative image, such as that received through the calotype process, and the visual acuity of the daguerreotype. It became apparent that transparent glass would be the most sensible medium to employ as a "base" on which to spread a light-sensitive emulsion. Unfortunately, the sensitizing silver salts would not adhere to the glass surface. Countless compounds were attempted and discarded in the search to solve the problem. One enterprising soul even attempted to mix the silver solution with the sticky slime exuded by moving snails.

In 1847, Claude Félix Abel Niépce de St. Victor, a cousin of Daguerre's partner, Joseph Nicéphore Niépce, was the first to succeed in adhering an emulsion to a glass plate using egg whites, or albumen. The resulting images were negatives, therefore reproducible, and had remarkably clear detail. Unfortunately, the low sensitivity of the emulsion necessitated very long exposures. This negated the possibility of using the albumen process for portraiture. However, the clarity produced was so far superior to the calotype that photographers specializing in immobile still lifes and travel views used albumen plates with great financial success.

Collodion—A Quick but Awkward Solution

Finally, in 1851, the process that ultimately solved the problem of lengthy exposure times was announced by an Englishman, Frederick Scott Archer. Archer had experimented with a substance called collodion (a combination of guncotton, alcohol, and ether) in order to repair some albumen glass plates that had been broken during handling. The collodion was transparent, membranous, and tough. The medical profession had used this liquid for years to cover the wounds of burn victims until the afflicted area could generate new skin. Archer discovered that by coating a piece of glass with a mixture of collodion and silver iodide emulsion, exposing the plate, and developing it immediately (before the emulsion dried), exposure times could be decreased dramatically. All of these actions took a bit of facility, as well as a convenient darkroom. Those who wished to use this process "in the field" had to carry a lightproof tent at all times. Even with the awkwardness of carrying camera, plates, chemistry, and the darkroom, this process was infinitely superior to the albumen plates. When the collodion emulsion was moist, the exposure time was decreased sufficiently to enable photographers to shoot any subject, including portraits. Archer graciously offered his process to the world without charge. However, when it was announced, Talbot was outraged, as he considered this new discovery to be an infringement of his calotype patent rights.

It is interesting to note that Talbot allowed his initial discovery to be announced with such freedom, for after issuance of the patent, he became almost maniacal about protecting his calotype process. Perhaps he had learned a few marketing ploys from Daguerre. Talbot may have seen the acclaim and financial rewards that Daguerre was reaping both from his annuity in France and from the sale of his licenses in England. He may have thought that such fame and reward should be his as well. Whatever the reason, when Talbot chose to grant licenses, they were very expensive. He charged any person who licenced his calotype process between £100 and £150 per year for the privilege. He was also very forceful in guarding against patent infringement. His actions were so energetic that they effectively stifled creativity in the medium.

Amateur and professional photographers alike thought Talbot too aggressive a protectionist and appealed to the Royal Academy and the Royal Society for relief. A partial easement was effected in 1852 when Talbot relinquished all control of his invention except its use for commercial portraiture. The easement was complete three years later. After much litigation and many hard feelings, Talbot was finally forced to relinquish his virtual stranglehold, on the life of photography in Britain, in 1855.

The Ambrotype—An Affordable Method

During this time, in America, another daguerreotypist was busily applying for patents using Talbot's and Archer's discoveries. James Ambrose Cutting realized the tremendous potential in the collodion process. Although the initial idea for the process was a direct attempt to produce a reprintable *negative* image, Cutting recognized that the fabrication of a unique *positive* image was also possible. These ambrotypes (derived from the Greek word "ambrotos," meaning immortal, imperishable) were underexposed or bleached collodion negatives that, when viewed on a dark background, appeared as a positive image.

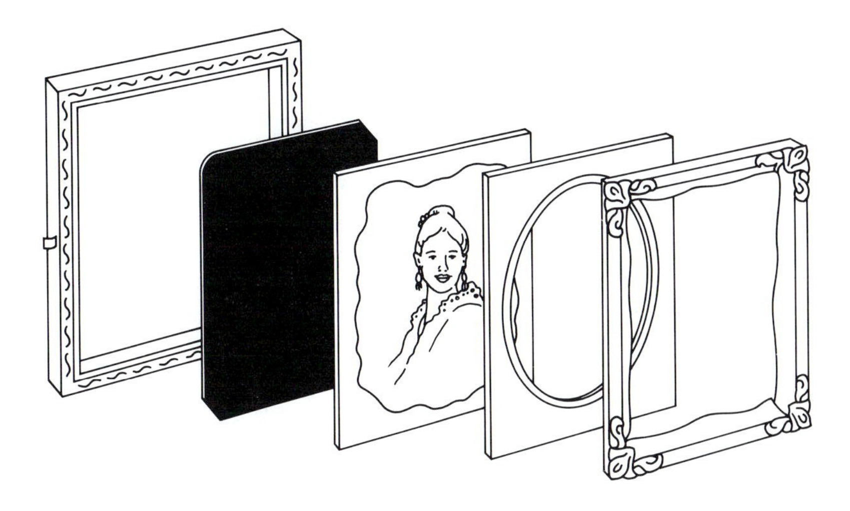

2-4
A disassembled ambrotype.

Cutting received three patents pertaining to the collodion process: two were for improvements to the emulsion; one was for the method used to glue the image between two plates of glass—a necessary step to protect the image from fingerprints and scratching. Thus safeguarded, the image was then backed with a dark substance. Sometimes lacquer paint was used, other times black paper or cloth were used. This "sandwich" was encased in a leather-backed wooden case, an oval frame was added, and a border of thin brass was placed on top to keep the elements together (Figure 2-4).

To the untrained eye, the resultant product looked remarkably like a daguerreotype. Although the image lacked the brilliance and mirrored quality when compared with a daguerreotype, the low cost of materials made it more profitable. When Cutting introduced his process in America, daguerreotypes were selling for between $1 and $5 apiece. A price war ensued between daguerreotypists and ambrotypists. Eventually, daguerreotypes were offered for as little as 50 cents. Ambrotypists, however, were able to offer portraits for 10 cents. Economics bested aesthetics, and the ambrotype began supplanting the daguerreotype.

2-5
unknown photographer
Unidentified Civil War Soldier
ca. 1862, tintype
Chicago Historical Society

As with most early photographic processes, ambrotypes had their share of problems. One problem was the time lapse of the emulsion. Even though Cutting had improved the sensitivity of the plate over Archer's original method, an exposure made with premixed collodion emulsion could take 20 seconds—too long for a comfortable portrait. To shorten the exposure time, the emulsions had to be very fresh. Also, the plate had to be coated, exposed, and developed while still in the wet–plate stage. For this reason, a cumbersome darkroom had to be immediately available. Also, Cutting (in keeping with Talbot's precedent) demanded hefty sums for licenses which would allow studios to process ambrotypes. The normal cost, in 1855, for permission to produce Cutting's process was $1,000. For more than a decade, Cutting vigorously prosecuted anyone using his process who lacked a license. However, in a curious bit of poetic justice, Cutting eventually lost these propietary battles in court. In 1868, a year after his death in an insane asylum, a U.S. Patent Office ruled that his heirs could not renew his claims to the silver bromide process (which had speeded the sensitivity of the emulsion) because the patent should have never been granted in the first place.

The fact that glass ambrotypes were less expensive to produce and sell than the equally fragile daguerreotypes, vaulted them briefly into the forefront of the photographic world. Their prominence lasted for less than ten years. In 1863, *The American Journal of Photography* reported that, "At the present time glass is less used for ambrotypes than iron plates. The name 'Ambrotype' may soon become obsolete."[3] The iron plates to which *The Journal* referred were the physical support for a process that, though using the collodion or "wet–plate" emulsion, was faster and less expensive than even the ambrotype. Also, because the support was metal, it was relatively impervious to rough handling.

The Tintype—Fast and Affordable

In 1856, Hamilton Smith, William Neff, and Peter Neff discovered that collodion emulsion could be poured onto any number of surfaces, including leather and cardboard. In one experiment,

Smith sensitized a preblackened sheet of thin iron, then exposed it in a camera. The idea worked and, soon, images on iron plates were sold to an increasingly enthusiastic public. Smith and Neff originally called them "melainotypes" for the Greek "melaino," meaning dark. An Ohio manufacturer preferred the term "ferrotype" from the Latin "ferrum," meaning iron. Eventually, they were simply referred to as "tintypes." As metal was a less expensive support medium than glass, prices for tintypes were a fraction of the cost of an ambrotype. Even the least affluent person could afford the price of a tintype.

The process of tintyping enjoyed an era of universal popularity for nearly eighty years. Tintypes were made on metal, so they were not fragile. As their introduction coincided with the start of the Civil War, soldiers could carry pictures of their loved ones into battle. A tintype did not require a protective case, so their bulk was lessened, thus increasing portability. They could also be made "instantly." A tintype could be coated, shot, developed, and into the hands of the customer in less than six minutes. Tintypes could be made in any size and by the 1860s they were popular as inserts in precast jewelry settings.

Most importantly, tintypes were affordable. The materials used in their production were readily available; no licenses needed to be purchased in order to produce them, leather and gilt cases gave way to cardboard frames, and the chemistry cost pennies. The expense of a tintype was further reduced with the use of a multilens camera, by which several images could be made at once and then cut with tin snips after development.

The Carte–de–Visite and Cabinet Photograph

As the tintype supplanted the daguerreotype and the ambrotype, the carte–de–visite (a further development of the collodion process) seriously eroded the market for tintypes. Patented in France by Andre Disdéri in 1854, its function was casual, similar to that of the tintype. As the name implies, cartes–de–visite were calling cards—mementos left with friends after visiting. The low cost and the ease by which these reproducible, portraits on paper could be fabricated, made them instantly successful. There were a few configurations of cameras used, although most employed a multilens system (similar to that used for tintypes), which could produce 8 to 24 photographs at a single sitting. Some cameras allowed for repositioning of the subject during the shoot. Others produced exactly the same pose many times on the same glass plate. Once the exposures had been completed, the plate was developed and printed. The paper positives were then cut into individual "frames" and pasted on light cardboard. A collection of these cards could then be carried in one's pocket or placed in specially manufactured albums. Its popularity can be surmised by a word coined especially for the subject—"cardomania."

Subject matter for cartes–de–visite soon surpassed the realm of personal calling cards. Theatrical personalities, historic sites, even the grotesque freaks at P.T. Barnum's circus were photographed and printed for mass distribution to an eagerly voyeuristic public. When Prince Albert (consort to Queen Victoria) died in 1861, 70,000 carte-de-visite portraits of his likeness were sold. Carte-de-visite images ranged in size from 3x5–inch postcards to those smaller than postage stamps.

2-6
André Disdéri
uncut print from a carte–de–visite negative
ca.1860
IMP/GEH

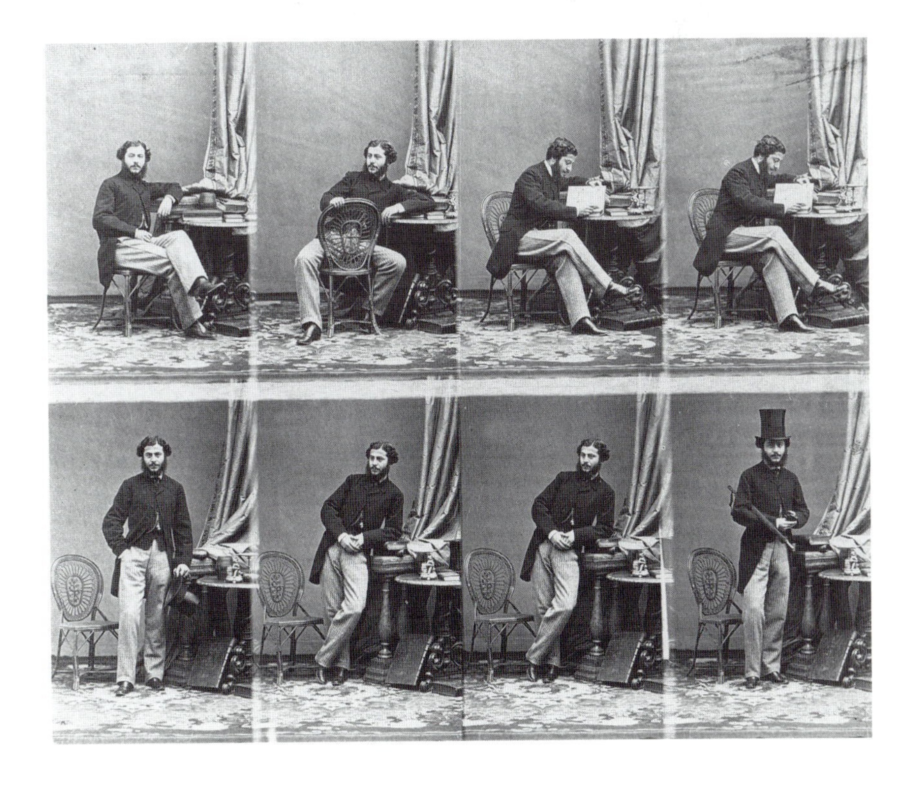

A few years later, larger and more formal "cartes" were made available. They were called "cabinet photographs" and were almost always used for portraiture. The backgrounds for these more formal images often utilized elaborate stage props to promote an idealized vision of the subject. Also, as the negatives taken were physically larger than those of the carte-de-visite, it was possible to retouch the image, thus removing moles and other unsightly blemishes.

Early Photographic Printing Methods

Once the problem of reproducible negatives had been solved using the collodion process on glass, technology was forced to find a suitable medium for printing. In 1850, the same albumen emulsion suggested by Niépce de Saint-Victor in 1847 (see p. 18) was used in the photographic printing process by Louis-Désiré Blanquart-Evrard. While the initial trials did not use developers and required a long time to produce, eventually Blanquart-Evrard contrived a paper that could be developed in gallic acid. The speed with which prints could be made using this paper enabled businesses to process 400 reproductions a day from each negative.

Other printing processes were being explored during the era of the albumen print. Between 1840 and 1858, research into emulsions made of gelatin, carbon black pigment, and potassium dichromate was quite successful. These products were called "carbon prints" after the color of the dried pigment used. Cyanotypes, or blueprints, were another photographic printing process utilized in the mid-1840s. Blueprinting was a simple and inexpensive process, but many photographers found the blue tint produced by the chemistry (formed through use of potassium ferricyanide and iron salts) too visually obtrusive.

Most photographic negatives were printed through the use of a contact frame and sunlight. Since the negative was in direct contact with the printing paper, the resultant image was exactly the same size as the negative. Although enlargers, as we know them, were not invented until years later, photographers occasionally used what they called "solar cameras" to make enlargements from their negatives. The optical system of these solar cameras is quite similar to that of a modern-day slide projector.

Exposures using solar camera enlargers took hours—sometimes days. Unskilled laborers and apprentices were hired for the sole purpose of keeping the solar camera in alignment with the sun's passage through the sky, thereby maximizing the sun's effectiveness. Unfortunately, the resulting prints were often of very poor quality. It was not until 20 years later, in the mid-1880s, with the invention of more sensitive printing paper, that the system of negative enlargement was proven commercially feasible.

The collodion (wet plate) process that had been introduced in 1851 allowed for nearly universal representation of a variety of subjects. However, as the process still required a cumbersome traveling laboratory, it was still too technically oriented for most people. To address this problem, various individuals experimented with new chemical mixtures to dispense with the need for a portable darkroom.

Between 1864 and 1867, collodio-bromide plates were introduced. They could be used while dry and could be processed at any point after exposure, but they were relatively insensitive to light. Notes from the manufacturers stated that these new dry plates required an average of three times the exposure time of wet plates. It was a trade-off. The dry plates were certainly more convenient, but images could be made only of immobile subject matter.

Gelatin Dry Plates—Quick and Convenient

In 1871, the *British Journal of Photography* published a letter from Richard Leach Maddox in which he described a gelatin emulsion that had the potential to solve this problem. Maddox, a physician, stated in his letter that he did not have the time to pursue the idea of gelatin dry plates and he offered it to any reader who would like to further the concept. Seven years later, after many modifications, prints were produced from previously manufactured negatives that had been exposed in a fraction of a second in normal sunlight and processed at the leisure of the photographer. An era so revolutionary had dawned that the *British Journal Photographic Almanac* printed a ditty mirroring the thoughts of every photographer:

"Onward still, and onward still it runs its sticky way,
And gelatine you're bound to
use if you mean to make things pay.
Collodion—slow old fogey!—
your palmy days have been,
You must give place in the future to the
plates of Gelatine."[4]

Manufacturers immediately began to produce gelatin plates by the thousands and the myraid problems that had plagued the photographer were solved. People were suddenly able to simply *take* pictures, not *make* them. They were released from the dark tent and minichemistry set that were necessary in which to stand to coat and process their negatives. Also, because of the sensitivity of the gelatin plates, the photographer no longer had to carry a tripod into the field. Gelatin film processed fast enough to allow for hand–held exposures. The invention of this emulsion and the elimination of the tripod ushered in a new era of photography. Finally, any interested individual could take his own photographs with a small financial outlay and absolutely no knowledge of chemistry.

Hand Held Cameras—Popular but Problematic

In 1888, George Eastman, was one person who recognized the many ramifications of this new discovery. He founded a company in Rochester, New York, that specialized in the manufacture of gelatin dry plates. He initially called the company The Eastman Dry Plate Company, then simply Eastman Kodak.

Some historians believe that Mr. Eastman conjured up the name Kodak, as it was easily pronounceable in all major languages. Others submit the idea that he appreciated the "firmness" represented in the letter K. Still others ascribe to the theory that George Eastman chose the moniker Kodak because it mimicked the sound made when the shutter button was pushed on one of his later inventions—the preloaded, hand–held, Kodak #1 camera.

The original Kodak camera measured 3¼ x 3¾ x 6½ inches. It was designed to be used with Kodak's previously invented "American film," which was roll film (actually paper) coated with a gelatino–bromide emulsion. By 1891, the paper backing had been replaced by transparent nirocellulose.

For $25, a camera could be purchased, already loaded with film sufficient for 100 shots. When all photographs had been taken, the owner shipped the camera back to the factory in Rochester. For a $10 charge, the film was developed, printed, backed, and a new roll of 100 exposures was reloaded into the camera. Thus, the advertising slogan, "You press the button, we do the rest," was quite literally true.

Although the advent of hand–held cameras, gelatin plates, and the simplicity of the Kodak system, made photography easier for an amateur, two stumbling blocks still had to be overcome. The first was the the absence of an aiming device on most hand–held cameras.

2-7
Necessary wet–plate photographic paraphernalia. The collodion, or wet–plate, emulsion required instant access to a darkroom (complete with tents, chemistry, trays and various other bulky tools) in order to produce photographic images.

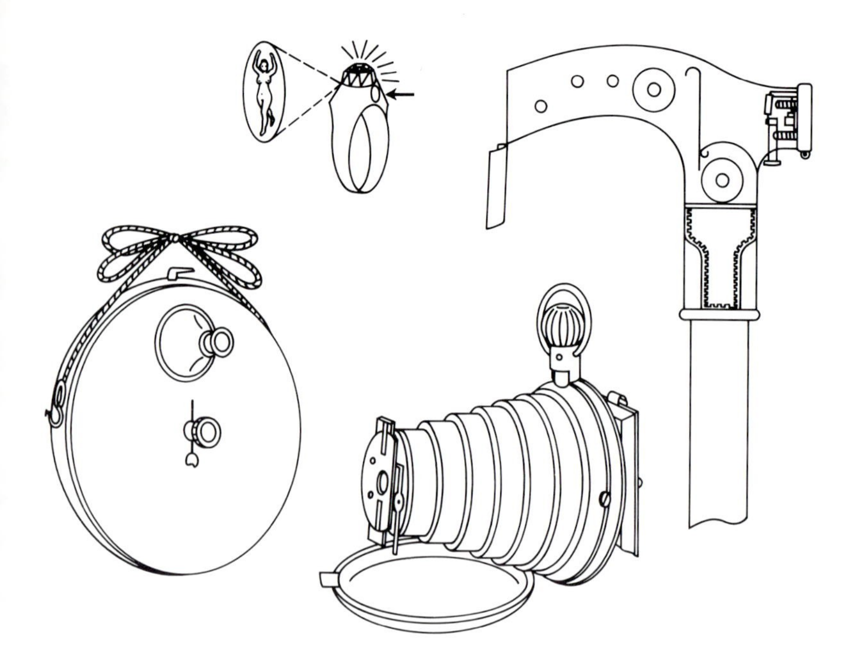

2-8
Miniature "detective" cameras. During the decade of the 1890s, concealed miniature cameras enjoyed a great deal of popularity. These machines were incorporated into walking canes, bowler hats, cravats, briefcases, or any number of ordinary objects. Negatives produced by these cameras were marginally useful, as precise framing was impossible and the film used was of a miniscule format.

Miniature, or "detective," cameras were manufactured soon after introduction of the first Kodak (Figure 2-8). These cameras either appeared as small rectangular boxes or were elaborately disguised within bowties, hats, luggage, even handguns. When these cameras were aimed at a subject, the point of reference was very broad because there was no device provided with which to view the subject directly. Often the photographer missed his subject entirely. Even Kodak's revolutionary camera came equipped only with a small arrow imprinted on the top of the camera body, presumably to prevent the photographer from taking a picture of his own nose. It is no wonder that, because of the inability to accurately direct the camera, photographers who used these small, portable cameras were called "snapshooters," a term used to describe hunters who shot quickly from the hip without aiming their rifles.

The second irksome problem of the photographic technology of the 1890s was inherent in the film, not the camera. Since the inception of photography there had been a persistent desire to represent the true range of black and white tonal values translated from our human optical experience of color. Wet–collodion plates and the early gelatin plates were overly sensitive to blue light and insensitive to the green, yellow, orange, and red rays of the spectrum. As a solution to this problem, Professor Hermann Wilhelm Vogel, in 1873, developed the system of "optical sensitizing." By adding dyes to existing photographic emulsions he was, in his words, "able to make bromide of silver sensitive to any color."[5] This remark was only partially true, as the film was still relatively insensitive to red light. He had developed what we today would term "orthochromatic film." It was not until 1906 that "panchromatic film," or film which was able to accurately translate all colors of the spectrum into a range of black and white tones, was introduced by Wratten and Wainwright. Although black and white films and emulsions continued to be improved in sensitivity and acuity after the early 1900s, the ability to record the chromatic shades and hues of color was still a dream.

Early Experiments with Color

A search started as early as the mid–1840s to find a method by which to record and retain accurate color on film. During the previous twenty years, it was noted that various tinges (mostly reddish) appeared on exposed film and daguerreotype plates. Although in 1850, Levi Hill (a New York minister), announced that he had succeeded in fixing colors on daguerreotype plates, it is doubtful that he had a great degree of success. In fact, a writer for *Humphrey's Journal of Photography* wrote in 1865 that Hill's process "was done by an accidental combination of chemicals which he could not, for the life of him, again produce."[6] Other respected photographers of the day accused Hill of outright trickery, suggesting that his results were actually daguerreotypes carefully colored by hand.

The experiments of the British physicist James Clerk Maxwell were far less questionable. In a dramatic performance before the Royal Institution of London in 1861, he proved that a crude but undeniable color image could be obtained by photographing and projecting three slides of the same object on the screen, and filtering them through three different optically colored solutions (Color Plate 1). Although it was a truly successful experiment, the reality of viewing color photographs in this manner was rather inconvenient. Although, in 1869, Du Haron and Cros made trichrome color prints, it was a very time–consuming operation. So, for nearly 30 years, between the 1860s and the 1890s, an audience had to accept the awkwardness of sitting in a darkened room, viewing three–plate preregistered color on a screen in order to experience color photography. It was not until 1893 that John Joly of Dublin took full color images contained on one photographic plate. Joly had placed a glass screen imprinted with translucent lines of red, green, and blue in front of a black and white glass plate. During exposure, the color divided itself into individual components. After a complicated development process and conversion from a negative to a positive, the result could be viewed as a color image if the original screen were removed and light were shone through it. Other, similar processes were attempted and were equally complex, time–consuming, and not altogether satisfactory.

Finally, in 1907, the Lumière brothers of France marketed the Autochrome process—the first relatively inexpensive, commercially feasible color photographic system (Color Plate 2). Like Joly's, the Lumière's color system produced a slide (or transparency). Unlike Joly's, it used a random dot pattern consisting of potato starch, rather than a mosaic, gridded pattern. Naturalists took to the process at once. Jacques Henri Lartigue embraced it in his work and stated that "life and color cannot be separated from each other."[7] Across the Atlantic, Alfred Stieglitz, champion of photography as an art form, received a supply of plates and shared their potential with colleagues (among them, Edward Steichen). He mounted a successful public exhibition of Autochromes at his New York "Little Galleries of the Photo–Secession."

Although popular, the Autochrome process, too, had its problems. Among them were the inability to produce enlargements; the possibility of unstable color; the requirement of a time–lapse exposure which took fifty times longer than comparable black and white film; and the inability to provide prints without exceedingly complicated procedures. Even though the drawbacks of the Autochrome were significant, the public penchant for photography representing life—in all its glorious colors—took precedence.

Although the general public did not mind the idea of a singular, nonreproducible slide, such as they would receive through the Autochrome system, they were also accustomed to the option of multiple printing from an individual *negative*. Many systems were tried in order to produce a color negative that would yield acceptable color prints. One suggestion required shooting three sheets of film simultaneously in a "one–shot" camera, each through an appropriate filter. When developed, the negatives could be printed by a variation of the carbon process (now called assembly printing) or by either the carbro or wash-off relief method (now called dye transfer) developed during the 1920s and 1930s. While the resultant full–color prints had a beautiful quality, the process was cumbersome and time–consuming.

Breakthroughs in Color Photography

It was not until 1935 that Eastman Kodak marketed a film that consisted of three color emulsions coated on a single piece of plastic film. Although it was a transparency film, the speed, sharpness, and clarity of color made it instantly successful. However, processing of the film required precise control. True to its initial marketing concepts, Kodak required that the exposed film be sent to the factory for development. The following year, a German company, Agfa, which had been a pioneer in the color "grainscreen" process, introduced Agfacolor Neu slide film. By 1942, both Kodak and a German company, Ansco, had devised films that a photographer could process himself. Concurrently, Kodak announced a negative–to–positive film, which they dubbed Kodacolor (Color Plate 3). Finally, in 1947, a photographer who wished to expose, develop, and print his or her own color imagery was able to do so using the Kodak Ektacolor system.

Another truly revolutionary system was introduced in 1947 by Edwin H. Land—Polaroid instant photography. This system answered a dictate demanded since the inception of the daguerreotype—that of speed. With the Polaroid system, a person could shoot and develop a photograph in a matter of minutes. No chemistry was necessary other than that provided in the film pack. No knowledge of camera workings was required other than that of focusing and shooting. Land chose The Optical Society of America's meeting on February 21, 1947, to demonstrate his process. The Society's program described it as "a new kind of photography as revolutionary as the transition from wet plates to daylight–loading film."

It was an odd choice of phrases, as the system was based on an idea virtually as old as photography—that of sensitizing, shooting, and developing film inside a restricted area.

Polaroid's spontaneity and convenience appealed immediately to a public that was—in the 1950s—increasingly involved with the concept of speed. MacDonald's was introducing fast food. Airplanes were quickly becoming the chosen mode of transportation. Supermarkets were supplanting the corner store and promoting the concept of one-stop shopping. Edwin Land's instant photography was a natural idea for the time. Within five years, more than one million Polaroid cameras were sold. Constant improvements were made on the camera system, as well as choice of available films.

In 1972, in keeping with increasing public concern for ecology, a completely waste-free Polaroid process was introduced: the SX-70 system (Color Plate 4). This process was a self-contained, completely automated single-lens-reflex system that could be shot and processed within minutes. Response, as expected, was tremendous. Eastman Kodak took note of these developments and quickly joined the field of instant color photography. Although their marketing strategies were different, with Polaroid aiming toward a sophisticated user and Kodak toward "snapshooters," the rivalry was so intense that Polaroid lodged ten separate patent infringement suits against Kodak within a few years. The court system found the lawsuits in favor of Polaroid, and Kodak was forced to cease production of its instant products.

Early Artificial Flash—Dangerous and Unpredictable

Although it had taken over a century to produce chemically instant images, the ability to capture immediate action had been achieved much faster. In 1851, Talbot used electric batteries in an attempt to increase illumination and shorten exposures. Thirteen years later, in 1864, an oxyhydrogen flame reflected against a wall of lime was used for the same purpose. That same year, magnesium wire was ignited to provide illumination for a photographic project inside coal mines in England.

Magnesium, a very common metal, was an excellent choice for instant illumination. It was available in powder, ribbon, or wire form, was easily ignited when dry, and burned with a short, blinding flame. Timothy O'Sullivan had used magnesium flares when photographing the Comstock Lode mines. In 1865, photographs were taken using ignited magnesium inside the Great Pyramid in Egypt. However, as with most new discoveries, it had a serious drawback: it was extremely dangerous. Unless the amount used was measured very carefully, the flare could explode, or at the very least, temporarily blind the photographer or his subject. Lesser effects included the emission of an acrid white smoke that could choke the photographer.

Flash from the 1880s to the 1930s—Significant Improvements

In 1887, two German inventors mixed a new compound in order to provide an instant, less dangerous photographic light source. They called it "Blitzlichtpulver," or flashlight powder. Magnesium was still a component, but it was cut with guncotton and placed in a metal tray, thus producing a safer and more consistent light. This mixture burned very quickly, with a minimum of smoke, and provided a quick and safe "flash" of light. Jacob Riis used this method of illumination while working on his photographic exposé of the living conditions in the Lower East Side of New York (Figure 7-2). He titled the layout, published by the New York *Sun* in 1888, "Flashes from the Slums."

A further improvement in instant illumination appeared in 1925 with the advent of the flashbulb, invented by Dr. Paul Vierkötter. By encasing magnesium wire in glass, instant artificial illumination finally was made safer and smoke-free. Low voltage electric current was used to ignite the bulb, thus eliminating the need for an open flame. Four years later, the amount of illumination provided by the flashbulbs was increased significantly by the inclusion of reflective foil within the glass tube. Further developments included the "multiple synchroflash," which made it possible to ignite several flashbulbs at once in order to create additional, aesthetically controlled light.

The invention of energy–storing capacitors and transistors made nondisposable, xenon–filled flash tubes possible. Initially called "speedlamps," they enabled amateur, professional, and scientific photographers to take pictures in low ambient light and permitted great latitude in the quest to freeze action.

In 1931, Dr. Harold Edgerton developed a variant to the concept of artificial illumination through single flashes. He accomplished this with stroboscopic light. "Strobes" emit flashes of light intermittently at very exact, extremely short intervals. Using this invention, Edgerton was able to produce images of the flow of motion and the trajectory of objects that move at speeds undetectable to the human eye. A curator at New York's Museum of Modern Art described them as "show[ing] empty spaces between the thinnest slices of time"[8] (Figure 12-13).

Looking Forward—Photographic Technology From 1960 to 1990

The knowledge gained from scientifically and technically oriented experimentation has expanded during the 1960s, 1970s, and 1980s. For example, holography—the science of giving images a third dimension—was perfected simultaneously in the Soviet Union and the United States in the early 1960s. Holographs are images that appear to actually exist in space, although the physical object represented does not exist where it visually appears. Split laser beams project the image of an object; interact with mirrors, lenses, and each other; and are then recorded on photographic emulsions to produce a holograph. Scientists use holography frequently, notably on archeological digs, not only to depict an object in three dimensions, but also to pinpoint a singular object in relation to other objects.

The advent of computers in the past 30 years has also given us dramatic new insights. Previously existing photographs can be "fed" into a computer through a digitization process and can later be retrieved and used for a variety of purposes. Three–dimensional modeling, or the showing of all contours of an object, permits both artistic and scientific study. Portions of digitized photographs can be recombined to form completely new images. Color computers are capable of working with both X–rays and telescopes to provide a wealth of information and beauty. Color–coded tomography can pinpoint health problems. Solar telescopes work with photographic film and color computers to elicit information on the heavens. While at this point, most applications are scientific, artists and journalists are increasingly using computer–generated imagery to enhance photographic vision.

In 1840, the world was astounded by the photographer's ability to create an object through the use of light. However, at that time, there was much confusion as to the best method of using this marvelous invention. One hundred fifty years later, these questions have been answered. We no longer marvel at photographs. We take them for granted. We use them for information and for aesthetic purposes. As society continues to evolve, so will the photographic image. Exposure times have changed from minutes—even hours—to 1 millionth of a second. Images have been transformed from bulky, moist, glass plates to objects that magically appear to suspend themselves in mid–air. In the future, photography will be not necessarily be governed by the tenets of the past. However, we may be certain that in the future, there will be further inventions, further problems, and, of course, further solutions.

We know what a photograph is...We have almost certainly taken a photograph, just as we have equally probably been part of one.
A photograph is (simply) an equation of light, time, and space.[9]
—Peter Turner

3-1
Louis Jacques Mandé Daguerre
ca. 1835, Diorama
IMP/GEH

3

Camera and Canvas—
The Alliance between Painters and Photographers

The painter sees nothing...but his opinion:
the camera has no opinions.[1]
—George Bernard Shaw

Before the advent of the camera, painting and sculpture were the two major forces in the world of fine arts. The two coexisted compatibly, as each represented a different dimensional mode. Paintings were normally two–dimensional works on canvas; sculpture represented all three dimensions. The nineteenth century invention of another two–dimensional art form, photography, skewed that arrangement considerably.

Sketching Methods Using Photo–Graphic Tools

Prior to the 1839 announcement of Daguerre's process, various photo–graphic (light–writing) devices were used as tools for artistic rendering. The camera obscura, for example, was used extensively by painters as a rather cumbersome "sketch pad." Painters found that the image formed by light on the interior wall of the camera obscura could be traced exactly as the exterior scene appeared (Figure 1-4). They particularly delighted in the sense of realism and perspective that the camera obscura afforded. From the tenth century to the early nineteenth century, painters used the camera obscura to assist them in their search for reality. Such artists as Canaletto (1697–1768) and Vermeer (1632–1675) were not hesitant to acknowledge the assistance rendered them by the camera obscura.

When, in 1807, W.H. Wollaston constructed the first camera lucida, portability was added to the repertoire of advantages photo–graphic devices afforded the painterly artist. Studies could now be made far afield through the use of this small table–and–prism system. A camera lucida superimposed, on paper, the image of objects at which it was pointed, allowing the "artist" simply to trace the image (Figure 1-7). Amateur painters found this new invention particularly satisfying, as it aided their untrained hand and made their sketches and paintings look more "professional."

Concurrent with the invention of the camera lucida (around 1800), an Englishman, Dr. William Lewis, was performing artistic experiments through the direct use of sunlight. Using silver nitrate, Dr. Lewis painted designs directly on semiporous white surfaces. Through exposure to the sun, images were formed on the material. Josiah Wedgwood, the British potter, acquired rights to use these methods (Chapter 2). His son, Thomas, was also interested in the process. Thomas painted images on glass, after which he attempted to transfer the image—again through the use of sunlight and silver salts—onto white surfaces. Presumably, he was interested in creating a photomechanical process that had the ability to endlessly reproduce existing drawings. Although he was partially successful in forming a negative image directly on the surface of an object, a process had not yet been developed that could arrest the action of the light. Therefore, once the image was formed, it had to be kept in almost complete darkness or it would fade to black. Still, these experiments represented progress in an alliance among drawing, painting, and photography.

Daguerre the Painter

While Wollaston was introducing the camera lucida as a means of assisting the painterly establishment, and as Wedgwood was attempting to form photographic reproductions of drawings, Louis Jacques Mandé Daguerre, inventor of the daguerreotype, was working as a successful painter in Paris. Originally apprenticed to an architect, then to a stage designer, Daguerre was fascinated by the arts. His oil paintings had been exhibited in the respected Paris Salon. However, Daguerre soon had bigger plans than simply creating normal canvasses of approved subject matter. He decided to enlarge his paintings and dabble in a new invention that was sweeping the Continent in the 1820s, the Diorama.

Essentially, the Diorama was, a cross between painting and public entertainment. Wall–sized scenes were painted on gauze and placed in specially prepared halls. Light was transfused through the gauze to create illusory effects. It is interesting to note that the Diorama used both paintings (made through the use of a photo–graphic device—the camera obscura) and the principles of light (essential to photography as well) to form its effect.

Daguerre's Diorama was inordinately popular in Paris. Many Parisians willingly paid the admission price in order to enjoy these fantasy landscapes (Figure 3-1). Its popularity continued until it burned to the ground, doubtlessly due to the painted and oiled gauze coming in contact with the illuminating flame. Its destruction occurred in 1839, the same year that daguerreotypes were introduced.

As mentioned in Chapter 2, the introduction of the daguerreotype process generated a great deal of interest and enthusiasm. However, a few years after the daguerreotype process had been perfected and acquired by the French government, Daguerre returned to working with canvas and palette. The paintings were, for the most part, perspective renderings of scenes that surrounded his estate in Bry–sur–Marne, France. Toward the end of his life, he spent very little time with his photographic invention.

Portraiture—Painters Versus Photographers

Although Daguerre's personal interest in his process waned after its introduction, the world continued its zealous support of his photographic discovery. The first popular use for daguerreotypes was portraiture, although initially there were many difficulties with this application. The lengthy exposure times were nearly unendurable. Also, because various contraptions were used to keep the subject immobile during a portrait sitting, the resulting image was stiff and unflattering. The subject looked lifeless and catatonic. In an attempt to remedy this problem, studios began to offer the services of "artists" (usually underemployed painters) who would tone areas of the plates with small amounts of pale color around the lips and cheeks. This practice made the subject appear more lifelike and therefore more "real."

The general public responded enthusiastically to these personal and inexpensive portraits. The painting establishment, in general, abhored them. Prior to 1840, expensive sketches and oils were the sole method of preserving a personal likeness. Painters could therefore look forward to a decent living if adept at capturing the features of a sitter. After 1839, with the advent of the daguerreotype, the assured livelihood of these portrait painters was jeopardized. Writers at the time were posing the possibility that artists would be driven to starvation when machines usurped their function. This concern was, of course, shared by the painting community. In a concerted move, the painting establishment attacked.

The painters employed every disparaging method imaginable to discredit the new process. They insisted that the daguerreotype could not be art, as it was made with a machine. Further, they added that the daguerreotype had other flaws. For example, it did not have the ability to reproduce colors. The painting establishment regarded the daguerreotype as simply a process of all chemistry with no sensitivity. In 1853, the painter Sir William J. Newton described the "seductive nature of the practice of Photography, and how it is calculated to divert him [the student of Art] from his principal object...."[2] M.A. Dwight, art critic, stated in 1856 that, "photography contributes...materially to the degeneracy of art."[3] In 1859, the French poet and art critic Charles Baudelaire summed up the feeling of the majority of the painting community by issuing a manifesto entitled "The Mirror of Art." In his article, he proclaims that "the photographic industry was the refuge of every would–be painter," that "the ill–applied developments of photography...have contributed much to the impoverishment of the French artistic genius," and ends his proclamation with a call for photography to "return to its true duty, which is to be the servant of the sciences and arts—but the very humble servant."[4]

The Photograph as a Visual Reference

It is odd that Baudelaire, after writing such an article, would have allowed himself to consort with a photographer, but he did. The photographer was Nadar. Born Gaspard Félix Tournachon, Nadar initially made his reputation in Paris as a caricaturist and painter in the bohemian Latin Quarter. When he decided to publish a series of large lithographs featuring satirical portraits of leading French citizens, he used a camera to capture their features. Baudelaire was one of his subjects, as were the painters Eugène Delacroix and Gustave Doré.

Delacroix and Doré were two of the small handful of painters who actively appreciated the role of photography in its relation to art. In fact, Delacroix wrote a friend in 1854 that, "it [photography] is the tangible demonstration of drawing from nature, of which we have had more than quite imperfect ideas."[5]

Delacroix was not the first to understand the vital role photography could play in painterly studies. In 1843 (four years after Daguerre's initial announcement), David Octavius Hill, a well-known Scottish painter, decided to create a large montage on canvas. His subject was to be a group portrait of the 457 people who seceded from the Church of England in order to form the Free Church of Scotland. Toward this end, he engaged the services of Robert Adamson, a professional photographer in Edinburgh. They were to travel the country photographing the clerics, then take the resultant images back to Hill's studio. Hill then planned to paint a "collaged canvas" from these photographic portraits.

Their collaboration proved very successful, although not terribly lengthy. Robert Adamson died five years after their partnership was formed. However, during the tenure of their collaboration, they accomplished a great deal. Their subject matter soon extended past that initial idea. Using the calotype process, they photographed fisherfolk, children, and nobility. The original canvas of Scottish clerics was finally finished by David Octavius Hill in 1866, twenty-three years after its start and four years before his own death.

Collaborations between painters, printmakers, and photographers thrived in the mid-1800s. In 1864, Matthew Brady's studio took the photograph of Abraham Lincoln that was ultimately etched on the American $5 bill. The Barbizon school espoused the *cliché-verre* process (Figure 12-7), which consisted of painters scratching designs into a coated photographic plate, then printing the resulting image. Edgar Degas and Toulouse Lautrec used photographs extensively as studies for later paintings. Some artists, such as Lautrec, Delacroix, and Hill, hired photographers specifically for certain projects. Others, such as Degas and Thomas Eakins, relied on existing photographs for reference. Still others had become so enamored of the visual truth inherent in photography that they had laid aside their easels and picked up cameras.

As early as 1840, Josiah Johnson Hawes, partner of Albert Sands Southworth (Chapter 6), was diligently working as a painter. As he wrote, "I practiced painting on ivory, likewise portraits in oil, landscapes, etc. with no teacher but my books." However, on seeing his first daguerreotypes, "[I] gave up painting and commenced daguerreotypy."[6] Both Henry Peach Robinson and Oscar Gustave Rejlander, champions of the combination print during the mid-1800s, had begun their artistic training in the painterly tradition. This is especially evident in H.P. Robinson's tome, *The Pictorial Effect in Photography* (1869), as he refers repeatedly to the painterly aesthetic of the day. Gustave Rejlander based his combination print, *Two Ways of Life*, on a fresco by Raphael (see Figure 12-5). Another painter of the late 1800s, Eugene Atget, had, by the 1890s, defected from the ranks of the painterly establishment to become the self-proclaimed "Official Photographer of Paris" (Chapter 10).

Most painters, however, were not interested in completely discarding their easels. Rather, they preferred to use the camera as a sketch pad. Others relied on published photographic images for reference. One book of photographs that became an invaluable reference source for painters was *Animal Locomotion,* by Eadweard Muybridge.

Photographic Assistance to Form-in-Motion

Born in England in 1830, Eadweard Muybridge journeyed to the United States as a young man and began to photograph the American West. His images of Yosemite won him a good deal of renown (Chapter 5), as did his invention of one of the first camera shutters. However, Muybridge's most famous work began as the result of—in his words—"an exceptionally

3-2
Eadweard Muybridge
"Daisy" Jumping a Hurdle, Saddled
plate 640 from *Animal Locomotion*
1887, photolithograph
MOMA

3-3
Eadweard Muybridge
from *Animal Locomotion*
1885, gelatin silver print
from master negative
IMP/GEH

felicitous alliance"[7] with California's ex–governor Leland Stanford. Stanford was a tremendously wealthy man who owned both a breeding ranch and a string of race horses. In 1872, Muybridge was contacted by Stanford and asked to settle a dispute among equine enthusiasts. This disagreement centered on the position of a horse's legs while at a full trot. During that time, painters and etchers drew "hobbyhorse" configurations to depict a horse's movement. Not certain that this was a correct representation, Stanford invited Muybridge to photograph his horse, Occident. Muybridge readily accepted the challenge.

Muybridge's initial experiments for Stanford failed due to the speed of the horse and the superhuman adeptness required to manually open and close a shutter in the short period of time required to freeze the movement. Soon, though, Muybridge fabricated a spring-tripped shutter system and captured an acceptable likeness of Occident in a full trot.

In 1877, after an enforced vacation from his work (having been tried and acquitted of murdering his wife's lover), Muybridge continued to photograph animals in motion. He hung a white background cloth against a long wall. He used a bank or battery of cameras, each fitted with a shutter that he claimed worked at less than $1/2000$ of a second. Strings attached to electric switches were stretched across the track and attached to the camera shutters. As the animal raced by, it snapped the strings and tripped the shutters.

After the publication of his images in *Scientific American* (October 1878) and subsequently in respected European journals, Muybridge was regarded as an international celebrity. Therefore, when Leland Stanford withdrew his support in 1883, Eadweard Muybridge was able to locate another sponsor. The University of Pennsylvania offered to provide Muybridge with complete financial support to continue his experimentation with capturing form–in–motion.

By 1884, Muybridge had extended his theme to include humans involved in every imaginable pose. He photographed people running, walking, skipping, pole vaulting, kicking bowler hats, wrestling, and dousing each other with pails of water. By 1886, Muybridge had fabricated many thousands of these images. The University selected nearly 800 for publication in a series of expensive volumes entitled *Animal Locomotion*. Later, Muybridge issued smaller, less costly, pamphlet–type editions, which he titled, *Animals in Motion* and *The Human Figure in Motion*.

Simultaneously, experiments similar to Muybridge's were being conducted in different parts of the world. Other photographers and scientists, notably in Europe, had been working simultaneously to devise methods of capturing form in motion. Etienne Jules Marey, after reading of Muybridge's early experiments, decided to explore the field as it related to his primary vocation as a physiologist. As his application of form–in–motion differed from those of Muybridge, he devised a distinctive technology of sequential images. He called his results *chronophotographs*, ("chrono" meaning number), in which multiple actions were recorded on the same piece of film.

Painters, sculptors, physiologists, writers, and physicians all demanded access to copies of these advances in photographic technology. The motion figure studies created a particular sensation among visual artists and the literary elite. The writer Emile Zola described the general feeling by stating, "In my opinion, you cannot say you have thoroughly seen anything until you have got a photograph of it, revealing a lot of points which otherwise would be unnoticed...."[8] The painter Edgar Degas used Muybridge's studies, among other sources. Degas, particularly, was so excited by the new frontiers of the medium that he became an ardent photographer. The sculptor Auguste Rodin, the painter Thomas Eakins, as well as the Cubists and Futurists of the early twentieth century all made use of, and gave appropriate credit to, the motion/figure studies created by stop–action photographs.

Photography Affords "Too Much Truth"

As might be expected, a portion of the painterly establishment, as well as a majority of art critics, disagreed with using photographs to assist painterly precision. Thomas Eakins, for example, was singled out at a meeting of the London Camera Club when Joseph Pennell (American illustrator) stated that, "If you photograph an object in motion, all feeling of motion is lost. This painter [Eakins] wished to show a drag coming along the road at a rapid trot...he drew and redrew....He then put on the drag. He drew every spoke in the wheels, and the whole affair looked as if it had been instantaneously petrified or arrested. There was no action in it. He then blurred the spokes, giving the drag the appearance of motion. The result was that it [the drag] seemed to be on the point of running right over the horses...."[9] This is just one example among many of a painting establishment that, at the time, eschewed photography as providing "too much truth."[10]

Of course, there were those painters who would not acknowledge the assistance that a photograph could afford their paintings. One of these was William Frith, who stated that, "In my opinion photography has not benefitted art at all."[11] However, in 1863, a London periodical noted that, "On a Derby Day, Mr. Frith employed his kind friend, Mr. Howlett, to photograph for him...and in this way the painter of that celebrated picture, the 'Derby Day' got many useful studies."[12] Other painters went even further to obscure the fact that they had received assistance from photographic studies. In the latter decades of the nineteenth century, for example, entire canvases were coated with emulsions, a photographic image was made, then it was painted over, obscuring the original photograph completely. In these cases, it is doubtful that credit was ever given to the person who tripped the shutter, or acknowledgement made of the original photograph.

Fortunately, these artistic fabrications were rare. In most cases, if photography were to be regarded as a reprehensible medium, it was done so in a forthright manner. Andre Derain, the French painter, exhibited a series of gaudy and dissonant canvases at the 1905 Autumn Salon in Paris. The work was tremendously abstract in terms of form and color. The term Fauve, which when translated means "wild beast," was used to describe both the canvases and the artists who fabricated the work. Henri Matisse, Georges

Braque, and Georges Rouault ascribed to this technique, which was believed to be an attempt to disassociate themselves from the Impressionist style, as well as from the influence of photography. Derain was particularly vitriolic, and was said to have remarked that, "It was the era of photography. This may have influenced us...against anything resembling a snapshot of life."

3-4
Henri Matisse
Le Dos
1914, pencil drawing
Bell Gallery, Brown University

Paintings and Photographs—Nonliteral Representation

In reacting against photography, the group of painters who called themselves Fauves were also turning away from the accepted ideal of painterly representation. Their demand for artistic freedom was echoed by the European Modernists and was notably championed by the American photographer Alfred Stieglitz. Three years before the Fauves exhibited *en masse* at the Paris Salon, Stieglitz had launched the Photo–Secession exhibition in New York (Chapter 8). The photo–secessionists were pledged to the same measure of visual freedom as the Fauves. Excellence in expression alone was stated as the primary factor in production of art works. The idea of perfect representation was disregarded.

This aesthetic credo was followed by many photographers of the late 1800s. Pictorialists of the time strove to shoot dreamlike images that mirrored the sensiblities of the painting establishment. Other photographers, such as Demachy, used the *cliché–verre* process, which would allow for partial photographic representation while other portions of the image would be less realistic and more expressive (Figure 12-7).

These ideals, espoused by Stieglitz, among others, shocked the photographic community of the early 1900s. For the 60 years prior to Stieglitz's Photo–Secession exhibition, the photographic community had applauded all inventions that promoted exactness in rendering "true life." Stieglitz was simply not interested in following those accepted aesthetic guidelines.

Upon opening his gallery at 291 Fifth Avenue in 1905, Stieglitz stated that his purpose was to show American as well as European photographs. He also mentioned that he was

3-5
Alvin Langdon Coburn
Vortograph #2
1917, gelatin silver print
MOMA

willing to exhibit "such other art–productions, other than photographic as the Council of the Photo–Secession will from time to time secure."[13] True to his word, in 1907, "291" hosted an exhibition of the drawings of Pamela Coleman Smith. In 1908, the two–dimensional works by sculptor Auguste Rodin and painter Henri Matisse were exhibited. In succeeding years, Stieglitz's alliance to the ideals of Modern Art seemed to eclipse his prior ties to the photographic community. He started to show Modern Art almost exclusively, with the justification that, "the Secession Idea was neither the servant nor the product of a medium. It is a spirit."[14] By 1912, his notion of "secession" had strengthened and broadened to become "a revolt against all authority in art, in fact, against all authority in everything."[15]

In the early 1900s, this revolt against the idea of "artistic authority" was evident in the work of painter Pablo Picasso. It was echoed in the canvases of the cubist movement. By 1914 and the advent of the First World War, the banner of artistic revolution was picked up by the European dadaists who believed that established traditions in art and literature were constricting and not representative of the society in which they lived. During the First World War, Europe was mired in unimaginable turmoil. It was left to the artist to represent that chaotic society. Painters such as Jean Arp (a founder of the dada movement) and Marcel Duchamp embraced the ideals of nonrepresentational art as a forum for revolution. A number of serious photographers, following Stieglitz's lead, agreed with this artistic philosophy. Their work evidenced this spirit.

In the United States, photographer Alvin Langdon Coburn, a member of the Photo–Secession, began abstracting the photographic image by distorting perspective. Ultimately, he completely removed any sense of reality when shooting his "vortographs." These images were the result of using his vortoscope, an optical device that resembled a kaleidoscope. In Europe, dada proponents in every medium were promoting visual confusion through the photographic process. Christian Schad, the German painter, exposed "found" objects such as rags and torn bus tickets onto photographic film to create his "Schadographs." Man Ray and Laszlo Moholy–Nagy created their photograms (Figures 10-3, and 12-8) at the German Bauhaus as a challenge to conventional thought and form. Painters were using photographs in their canvases; photographers were applying paint to their imagery.

This breakdown of traditional media—or new alliance between media—became especially evident in the collages and montages produced in Europe after the First World War. Such artists as John Heartfield (Figure I-3), Alexander Rodchenko, and Raoul Hausmann (editor of a dada journal) designated these images made from torn parts of other photographs as "photomontage." To them, the montages seemed to reflect "the chaos of war and revolution."[16] They also were used, with increasing regularity, as social and political commentaries.

The practice of expressing social commentaries through montages, paintings, and photographs was certainly not new, nor restricted to any particular national orientation. The new society formed in the 1920s as a direct result of the First World War and the second Industrial Revolution was one based on ideals that directly affected "the common man." In the Soviet Union, artists sought to awaken working class viewers to the meanings of Socialist orthodoxy by using photographs and text.

Photographs and Paint—Sharing the Same Surface

Later, in the 1950s and 1960s, the painter Robert Rauschenberg combined derelict objects with silk–screened photographs and paint (Color Plate 5). In these "neo–Dada" assemblages, he effectively communicated ideas about a rapidly changing world. Television and other media were, at the time, just beginning to alter the collective consciousness. Rauschenberg reacted to this new technological climate through his use of paint and photographs.

Andy Warhol, noted Pop artist of the 1960s and 1970s, also used photographs and paint to create imagery that directly reflected his view of Western culture. Warhol, in his canvases, used

3-6
Chuck Close
Maquette for Alex Katz
1987, color transparency
©Chuck Close
courtesy Pace/MacGill Gallery

3-7
Chuck Close
Alex
1987, oil painting
©Chuck Close
courtesy Pace/MacGill Gallery
(original in color)

current photojournalistic images that he photographically silk–screened in a repeat mode. These were intended to confront us with the issues of our lives, as well as reflect the increasingly mechanized society in which we live. During the 1970s and 1980s, Warhol personally shot most of the photographs he used, as he liked to think of himself and his camera as integrated machines. In his words, "The reason I am painting this way is that I want to be a machine and I feel that whatever I do and do machine–like is what I want to do."[17] Warhol often used that most mechanized of twentieth century cameras, the Polaroid SX–70 (Chapter 2). Likewise, contemporary painter David Hockney uses an SX–70 to shoot multi–image collages of portraits and landscapes.

Painters and Photographers—A True Collaboration

The Super Realist Movement of the late 1960s, and 1970s promoted this ideal of machine–like precision within the arts. This artistic style was foreshadowed in the 1920s and 1930s by the Precisionist Movement. One precisionist, painter Charles Sheeler, made very exacting photographic studies of modern architectural details. He then painstakingly translated the silver image onto canvas. Painters in more recent decades, such as Chuck Close, have chosen to "grid" photographic images and then paint them, as precisely as possible, onto the canvas. Recently, Close has rested his palette and taken up a Polaroid camera to make mural–sized assemblage portraits. Painter Audrey Flack uses a slide projector in her work. She projects photographic transparencies onto canvases tacked to a wall. She traces, then paints the original photographic image. Her methodology is a twentieth century mirror of that fifteenth century visual tool—the camera obscura.

Since the 1960s, some photographic artists have extended the photographic ethic to include application of paint to the silver surface of their photographic prints. Christopher James applies lightly colored enamels to the surface of his black and white photographic images. William Parker's male nudes of the 1980s are surrounded by heavy applications of opaque coloration. Contemporary artists such as Naomi Savage, Betty Hahn, and Todd Walker use various methods of coloration to embrace a spectrum of artistic photographic styles (Chapter 12).

In the 1980s, with the advent of computer–generated imagery and digitization of existing imagery, the aesthetic worlds of painting and photography realize a portion of completely common ground. Painters, since the invention of the medium, have enjoyed the freedom to selectively compose from "life" through the addition or deletion of subject matter or color. With a few notable exceptions, however, photographers have not historically had those choices. The new computer systems of the 1980s and 1990s now allow photographers the option to radically alter existing imagery in terms of both coloration and composition. Thus, photographic art has come to enjoy the freedom historically reserved for the palette and canvas.

From the camera obscura to the computer terminal, from the photograph as a notation for an ultimate painting to the photograph as a respected art form, the world's aesthetic perception has been profoundly altered due to "the magic black box." Paint on photographs and photographs in paintings represent an alliance between two media that has developed and strengthened the vision of art and our visual culture.

...the modern school of painting and photography are at one; their aims are similar, their principles are rational, and they link one into the other; and will in time, I feel confident, walk hand in hand....[18]
—P. H. Emerson
1886

4-1
John Thomson
Street Doctor
1877, carbon print, woodbury type, from *Street Life in London*
Museum of Art, RISD

4

Social Documents—Photography and Social Culture

It is not sociologists who provide insights but photographers of our sort who are observers at the very center of their times. I have always felt strongly that this was the photographer's true vocation.[1]
—Brassaï

4-2
W. Roberts
Street Seller of Birds' Nests
ca. 1850, wood engraving
after a daguerreotype
by Richard Beard
from *London Labour and London Poor*
New York Public Library/Astor,
Lenox and Tilden Foundations

Image makers who concentrate on social culture often reveal their personal feelings and prejudices. Their work can include a variety of orientations. It can reflect a romantic mode or act as a tool to redress strident social injustice.

Social culture photographs need not be aesthetically or technically "perfect." Firm foundations in classical composition are often disregarded. In fact, many social culture photographs have a rough, unfinished quality—as if the photographer were stating that social concerns took precedence over aesthetics.

Although social culture photographs can manifest such wide technical and aesthetic latitudes, all of them share two basic characteristics: a humanitarian point of view and a basis in fact. People are ordinarily the subject of social culture imagery. For example, when railroads were first constructed to span the United States in the mid–1800s, many photographers trained their lenses on the train cars, engines, or tracks, in order to glorify this new technology. The social culture photographer, however, sought out the railroad workers. He specifically captured these laborers—in groups or singly—at work or idle.

In addition to the humanitarian aspect, images of social culture are normally straightforward and unmanipulated. As they are meant to represent fact, they must be unretouched. This presented a problem for the first social culture photographers, as a method for precise photographic duplication had not yet been invented. The calotype existed, but the fuzziness caused by integral paper fibers lent an air of unreality to the images.

Laborers and Living Conditions—1850 to 1905

When Henry Mayhew and Richard Beard attempted to present the first social culture project, *London Labour and London Poor* (1850), it met with a generally negative reaction. Their thesis was meant to be an exposé of the living and working conditions of the London poor.

Mayhew worked diligently on the text and hired Beard to enliven the work with visual illustrations. Beard supervised the shooting of many daguerreotypes. However, as daguerreotypes were not reproducible, he was forced to give the plates to a wood engraver. The resulting illustrations were far from a precise representation of the original images. The engravings showed lifeless figures, standing stiffly against sketchy backgrounds. Because of this mediocre "artistic" translation, the public (whom Mayhew and Beard wished to incite) viewed the volume with a detached skepticism.

A later work, *Street Life in London*, was more successful in representing a portion of the English social culture as it existed in the late 1870s. John Thomson and Adolphe Smith conceived of a serial publication that would include written vignettes (supplied by Smith), accompanied by woodburytypes directed by Thomson. The woodburytype process was appropriately chosen for this task, as it was a photomechanical transfer process. Thus, woodburytype duplication of photographs provided a more acceptable, more accurate representation of the original image. In the preface to the book, Thomson noted "the precision of photography in illustration of our subject. The unquestionable accuracy of this testimony will enable us to present true types of the London Poor and shield us from the accusation of either underrating or exaggerating individual peculiarities of appearance."[2]

Mayhew and Beard, in their 1850 volume, *London Labour and London Poor,* had been criticized for overstating the poverty in the London slums.

This reaction was largely due to the lack of decent illustrative evidence. Twenty years later, Thomson and Smith's project, received a better reception, as they presented a "true" photographic representation of the abject poverty in their society.

A Danish man, transplanted to the United States in 1870, confronted the same problem as Beard and solved it in the manner of Thomson. Jacob Riis arrived in the United States in the late 1800s, just prior to the flood of immigrants from Eastern and Southern Europe. The housing in New York City, where most of these newcomers were forced to settle, was wholly inadequate for the sheer numbers of people arriving in America. Basements, even streetcorners, became homes for these foreign–born individuals who had emigrated hoping for a better way of life.

Riis, as a reporter for the *New York Herald,* was assigned the area where many of these immigrants lived. He was stunned by the squalid accommodations and wrote tremendously descriptive articles for the newspaper. However, these columns were greeted with the same skepticism that Mayhew and Beard had encountered. It was only when Riis presented photographic evidence of the unhealthy and overcrowded housing situation through slide lectures and his publication, *How the Other Half Lives: Studies Among the Tenements of New York* (1890), that the general public took notice and implemented programs for reform (Figure 7-2).

Since much of Riis' work was designed for the print media, his concerns are covered more fully in Chapter 7—Photojournalism. However, as his images have such a humanitarian concern, he can also be considered a social culture photographer.

This is also true of another early social culture photographer, Lewis W. Hine. Hine's photographs of labor conditions in the early 1900s depict a particular social culture. He photographed his subjects first as victims of their society, then as evidence for the print media. His most serious concern was the plight of small children trapped in the work force during the early period of American industrialization. The photographs of these children in squalid workplaces are among his most moving images. However, he did not confine himself to photographing children. People of all ages, caught in the trap of working long hours for very low wages, were fodder for Hine's lens.

Born in Wisconsin, Hine studied at the University of Chicago under a well–respected educator, Frank Manny. When Manny moved to New York for appointment as Superintendent of the Ethical Culture School, Hine followed. Hine was eventually employed to teach botany and nature studies at the school. In 1903, Manny placed a camera in Hine's hand and suggested it as a useful teaching tool. Manny did not foresee that Hine would take his suggestion so seriously. Hine not only taught his students through the use of a camera, he eventually taught all Americans about the life of industrial immigrants during the mechanical revolution of the early twentieth century.

Having come from a small town, Hine was astounded and amazed by the bustle, the size, and the burgeoning population of America's most thriving city. The population explosion, which occurred during the early 1900s, was primarily due to the number of immigrants flooding into New York's Ellis Island. In 1907 alone, over one million foreigners filtered through this American gateway. These people came to America for freedom and a new life. What many found was a new kind of slavery—working in sweatshops at wages that forced them to remain economically and socially stagnant.

In 1904, Hine began shooting his first series of images at Ellis Island. He photographed the faces, the paltry luggage, and the "processing" of the immigrants. Eventually, Hine followed these foreigners to their newfound homes, to capture on film their living and working conditions. After photographing this general subject for a number of years, he realized that in order to effect social change, he would have to affect those who viewed his imagery. To this end, he narrowed his focus and began to concentrate on photographing the plight of the immigrant child.

America in the early 1900s seemed to center its consciousness around the child. Prohibitionists aimed their anti–alcohol propaganda at the potential for destruction of the children of America. The beginnings of the revolution in

4-3
Lewis Wickes Hine
Child Labor in Georgia Cotton Mill
1909, gelatin silver print
Museum of Art, RISD

4-4
Lewis Wickes Hine
Child in Carolina Cotton Mill
1908, gelatin silver print on masonite
MOMA

elementary school curriculum transpired at this time. However, what should have been a forward movement for the betterment of *all* children actually centered only on the middle and upper class strata of the country. Many of the children of immigrant workers knew nothing of school. Out of financial necessity, some parents were forced to keep their children home. The youngsters would perform such menial tasks as picking nutmeats, stitching dresses at a piecework rate of pay, or working in mines.

A battle was being waged between the large industrialists (who insisted that it would be most beneficial for a small child to work at a bobbin machine than engage in "fruitless play") and Lewis W. Hine. In 1907, Hine was appointed to work as an investigator, photographer, and reporter for the National Child Labor Committee. For the next ten years, he traveled the country photographing youngsters in mines, mills, and in their homes. He used assorted ruses to gain entrance to these workplaces, occasionally reporting to a site disguised as a laborer, his camera concealed in his lunch pail. He often shot these children against the background of the immense machinery they operated in order to underscore the diminutiveness of his subjects.

In 1916, a portfolio of Hine's work was published using the halftone process. It took the country by storm. The general public was shocked by the conditions evident in Hine's imagery. A consensus was soon reached that resulted in federal legislation. The new laws encompassed rules concerning child labor and improved public education. Through work by the National Child Labor Committee, the newly formed labor union system, and the photojournalistic efforts of Lewis W. Hine, the future of millions of Americans had been irrevocably, thankfully, altered.

The Social Culture of the Workplace

The lives of the poor and their efforts in the workplace were continuously monitored by social culture photographers. The mid–1800s produced *London Labour and London Poor,* as well as the work of David Octavius Hill and Robert Adamson. Later work by Jacob Riis, Lewis W. Hine, and Peter Henry Emerson continued to highlight the lives of those who worked hard for little recompense.

As early as 1845, Hill and Adamson made calotypes depicting the locale and fishermen of the small town of Newhaven, Scotland. The project was initiated to publicize the inadequate fishing boats and tackle that needlessly endangered the lives of local fisherpeople. The images were softly lit, beautifully composed, and met with such strong response when shown to more cosmopolitan Scots, that funds were raised to correct the deficiencies in the fishing industry of Newhaven.

Within five years (1850), photographs appeared that depicted all manner of work forces and workplaces. The first were daguerreotypes of farm workers and miners, both in the United States and Europe. Soon after 1850, photographs of the new railroads spanning the United States captured the public's imagination. Then, as the California gold rush became a topic of international news, images of panners, their accouterments, and their styles of life were distributed to a curious public. As mining became more established, such renowned photographers as Timothy O'Sullivan descended into the subterranean tunnels to capture silver miners at work. Most of these photographs, taken between 1850 and 1880, downplayed any particular compositional aesthetic. The person at work was the focus of the photograph.

In 1885, Peter Henry Emerson (later recognized for his treatise, *Naturalistic Photography*) embarked on his photographic career by documenting fellow Englishmen in the area of East Anglia. Emerson generally treated his subjects with a great deal of aesthetic care. These humanistic photographs show a bygone era with much veracity. He photographed the marshes and their inhabitants—mostly poor farmers, fisherpeople, and laborers—for more than five years.

Social and Cultural Uses of Photography

Almost since its inception, photography was perceived as a method to understand different cultures or subcultures within a given society. Occasionally, these images would have a serious anthropological orientation, such as the work of August Sander (Figure 6-12). More often, however, popular interest was aroused by simple scenic images of "foreign lands" (Figure 5-3). As early as the mid–1850s, photographers made substantial livings traveling the globe, photographing countries and peoples who were considered "ethnic" or "foreign" to a Western audience. Starting around 1860, these glimpses

4-5
Peter Henry Emerson
Setting up the Bow–net
1886, platinotype
Museum of Art, RISD

4-6
Edward Curtis
Untitled
ca. 1900, photogravure
private collection

of foreign social cultures were available as stereographs. As discussed in Chapter 12, stereographs are two separate photographs, made from a perspective of approximately 2.5 inches apart, attached side by side on the same piece of cardboard. They were placed in a hand–held viewing apparatus called the stereoscope (Figures 12-2, 12-3, and 12-4). The resulting image appeared three–dimensional and quite true to life. Stereoscopes were a popular form of entertainment between 1860 and the turn of the century. They were considered essential items for every drawing room. During an evening's relaxation, families and friends would gather to marvel at the tribesmen of darkest Africa or the milkmaid of Austria.

However, not all social culture photographers were interested simply in a popular though somewhat frivolous depiction of different people. These more sociologically and anthropologically oriented image makers considered the camera as a method to capture the customs and lifestyles of vanishing cultures. From 1873 to 1874, John Thomson of England (who later worked with Adolphe Smith on *Street Life in London*) published 200 photographs entitled *Illustrations of China and Its People.* The series was the culmination of his four–year visit to that country and was designed to make "foreign" culture understandable to his fellow Britons.

During the 1880s, another Englishman, Sir Benjamin Stone, wanted to generate a "record of ancient customs, which still linger in remote villages"[3] of England. He envisioned the project as a means for future generations to better understand the history of their own culture. Stone envisioned it as a national project. He urged amateurs in all parts of England to conduct small photographic surveys in their hometowns—of their ceremonies, their landscapes, their singular customs. In 1897, he founded the National Photographic Record Association, in order to preserve this information for future generations.

In America, concurrently, photographers were both "snapshooting" their personal social cultures and also involving themselves with more weighty projects. The Native American of the Southwest was, due to the migration of the "white man," integrating and evolving into a very different culture. The old customs, the ceremonial attitudes of life, were eroding into the twentieth century. Adam Clark Vroman, a bookseller in Los Angeles, California, left his business in order to photograph this vanishing way of life. In his images, he attempted to depict the positive aspects of tribal customs. While he concentrated on the beauty and dignity of the Native American, he did not overromanticize the hardships facing Indian society at the time.

Edward Curtis of Seattle, Washington, also recognized the need to photograph the American Indian before its tribal purity disappeared into a more modern civilization. Curtis had been a successful portrait photographer before embarking on the project that would involve thirty years of his life (1907–1937). Assisted financially by banker and philanthropist John Pierpont Morgan, Curtis traveled to a multitude of Indian outposts, making carefully composed portraits of the different tribes and their ceremonial costumes. He also highlighted other portions of their social culture, such as tools, weapons, and housing. Curtis's images tended to be more posed and stiff than those of Vroman. Occasionally, their authenticity was somewhat suspect. For example, Curtis would sometimes have tribesmen dress in very old ceremonial robes in an attempt to revive a generational past. However, in an historic context, Curtis's motives were not necessarily intended to deceive the viewer. As Curtis also had somewhat commercial motives, twenty volumes and twenty portfolios of these images were ultimately published, under the title, *The North American Indian*. A foreword was written to accompany this work by Theodore Roosevelt, then President of the United States.

Culture and Subculture

The Native American was one of a number of ethnic groups who drew the attention of social culture photographers around 1900. Arnold Genthe, a German émigré, photographed San Francisco's booming Chinatown as early as 1895. Using a concealed hand camera, Genthe prowled an area of the California Bay city that few white people had ever visited. On these forays into the generally inhospitable "Canton of the West," Genthe captured images of a firmly entrenched foreign community existing within the confines of the "normal" American culture. The photographs have a rough, documentary style. Whether the subject is a father and daughter hastening down the street or a stuporous opium addict, the feeling evoked by the photograph is gritty and real.

E.J. Bellocq depicted another American subculture of the early 1900s. His work described the social culture of the New Orleans, Louisiana, prostitute. Bellocq was a little–known commercial photographer living in New Orleans around the turn of the century. The few sketchy reports that exist describe him as an odd man, in appearance as well as humor. His unsociable behavior matched his unremarkable commercial photographic skill;

4-7
E.J. Bellocq
New Orleans
ca. 1912,
silver print on printing–out paper
©Lee Friedlander, New City, NY
Museum of Art, RISD

4-8
Weegee
Murder in Hell's Kitchen
ca.1940, gelatin silver print
courtesy Wilma Wilcox
IMP/GEH

and Bellocq would have remained completely obscure in the history of the medium had not a curious event occurred. After his death in 1949, an acquaintance was cleaning out his personal effects. In a desk drawer this person discovered 100 glass plates generated by Bellocq. These plates depicted the lives of New Orleans prostitutes in various costumes and postures. They were always posed, although the lighting and attitude of the subject was casual.

Of course, as Bellocq was a professional photographer, it is possible that the images were part of a commission. However, it is more likely that Bellocq led a double life—antisocial and anonymous by day, voyeuristic photographer by night. Unlike some previous social culture photographers, such as Hine or Riis, Bellocq did not present the *demimonde* as sordid. Rather, his images are ones of beauty and peace, although they reveal a shadow culture.

Bellocq's photographs of prostitutes in New Orleans depicted working people and their environment. The images also went a step further in that they presented a largely unfamiliar subculture. Even today, the images fascinate viewers because of our innate curiosity (some may say voyeurism) concerning cultures at variance with ours—or hidden within our own.

Bellocq's portraits of prostitutes are the only body of his work to have survived the decades since his death. Fifteen years after they were unearthed in his possessions, they were shown to photographer Lee Friedlander, who ultimately bought them. Friedlander then printed the negatives, using an antique process that was historically accurate to the dating of the images.

With the advent of the hand camera, more photographers worldwide felt compelled to visually document areas of the culture that were not well-known by the general public. In the 1920s, the Hungarian, Gyula Halász, armed with a 6.5 x 9-cm hand camera, set out to photograph the underside of Paris nightlife.

Trained as a painter in Budapest and Berlin, Halász arrived in the City of Lights in 1923. Mesmerized by the metamorphosis of the city when the sun set, he began to photograph it, particularly at its most disreputable levels. During this project, he adopted a new moniker—Brassaï. It was a derivation of Brasso, his Hungarian hometown (which, by the way, was also the home of the legendary Count Dracula). Thus, armed with a new name and a camera that was portable for all situations, he took to the streets. As he says, "for five or six years, from 1924 on, I led a nocturnal life in Paris, sleeping by day and walking by night."[4] These images of pimps, prostitutes, pool halls and pleasure seekers were eventually collated into a book, published in 1933, and titled *Paris de Nuit (Paris by Night)*.

An interesting comparison can be made between the images captured by Brassaï of the social subculture of Paris and those photographed by American Arthur Fellig. Contemporaries, they were drawn to photograph urban cultures after dark. Like Brassaï, Fellig took a new, one–word/two–syllable name as he began his project to photograph the underside of New York City nightlife. Fellig chose the name "Weegee," after a board game popular at the time, the Ouija Board.

Many people who knew Fellig, or Weegee, thought he had chosen an appropriate pseudonym. They believed he was supernatural in his ability to be at a particular location at the appropriate time in order to capture the optimal image. More likely, it was not any kind of special power, but the police radio he kept in his car, that allowed Fellig to arrive at a location with split–second timing. Like Brassaï, Weegee published a compendium of his images, in 1936. Weegee titled his volume *Naked City*.

Jacques Henri Lartigue—
A Privileged Social Culture

Of course, not all social culture photographers of the period 1910–1935 were solely interested in capturing the sordid aspects of society's sub-culture. Some, such as Jacques Henri Lartigue, were content to photograph a more intimate, infinitely smaller subculture—that of the privileged class.

Lartigue was born in 1894. At the age of seven, his father gave him a hand camera. Although generally considered a portraitist (Chapter 6), as he photographed his large and wealthy family, Lartigue may also be considered a social landscapist, as he also made images of elegant Parisians (ca. 1910–1930). Lartigue kept a complete written diary of the times and his experiences. In 1938, he wrote, "People talk about decadence, they say that this is an era spoiled by self–indulgence and excess. And yet...I think we often judge a period by a rather small number of the people in it."[5]

Lartigue photographed a small number of subjects, yet his marvelously simple images—of family, of friends, of strangers—elegantly capture a brief moment in a changing social culture. Eventually, Lartigue's work was compiled into a book, *Diary of a Century*. In the afterword of this 1970 volume, photographer Richard Avedon describes Lartigue's contribution to the visual catalogue of that privileged French subculture: "It's staggering how lost (it) all is...to the whole world. Lartigue has shown us a laughter that is past and the laughter we have traded it for. He has shown us leisure as an adventure and as an indulgence and made us know the full impact of what is lost."[6]

4-9
Jacques Henri Lartigue
Avenue du Acacias, Paris
1911, gelatin silver print
©Jacques Henri Lartigue/ Photo Researchers Inc.

4-10
James VanDerZee
Couple, Harlem
1932, gelatin silver print
Museum of Art, RISD

An Emerging Black Subculture—New York City—1920 to 1950

In the early 1900s, James VanDerZee, a black American photographer, captured a specific climate within a portion of the American urban scene. He photographed the moderate–to upper–income black society that was then emerging in New York City. In 1916, VanDerZee opened a professional portrait studio in Harlem. Within a few years, he was sufficiently successful to leave the studio in order to photograph events in his community. Like Genthe's photographs of California's Chinatown, these visual legacies are part of the scant remnants of a powerful social structure that had emerged within the United States.

Between 1932 and 1950, Aaron Siskind, a white photographer and member of the Photo League, continued in this vein by working on a project called the Harlem Document. In this body of photographs, Siskind portrays the same physical landscape as VanDerZee, albeit a few decades later. The elapsed time, and perhaps the differences in races and concerns of the photographers, resulted in differing social landscapes within the same subculture. VanDerZee concentrated on portraying upscale members of the black community and their community events. Siskind, as a member of the Photo League, photographed a broader cross section of life in the black community.

The Photo League was formed by a group of politically conscious photographers. These photographers were committed to in–depth documentation of the particularly urban social culture. Founded by Sid Grossman and Sol Libsohn, its members included such photographers as Morris Engel, Jack Manning, Walter Rosenblum, and W. Eugene Smith.

4-11
Aaron Siskind
Untitled—from *Harlem Document*
ca. 1947, gelatin silver print
Museum of Art, RISD

The Social Culture of Economic Change—America—1929 to 1939

During, and just prior to, this period, the imagination, concerns, and economy of America were riveted on the Great Depression. This economic catastrophe, which started in 1929 and lasted for a decade, affected urban and country dwellers alike. As a measure to assist those in need, the United States Government formed the

Resettlement Administration in 1935. The concept behind the Resettlement Administration (RA), in part, involved the resettlement of the urban and rural poor onto communal farms or into suburban communities for training. The goal of this training was geared toward promoting self–sufficiency.

The Resettlement Administration, after a few administrative bumps, eventually evolved into the Farm Security Administration (FSA). Headed by Rexford Tugwell, the FSA was committed to many goals. One of these was the documentation of the project and the people it was geared to assist. To this end, Tugwell hired Roy Stryker to manage a cadre of photographers whose specific task was to "...introduce Americans to America...to help connect one generation's image of itself with the reality of its own time in history."[7] Through the work of such photographers as Walker Evans, Dorthea Lange, Ben Shahn, and Russell Lee, among others, a portrait of America during one of its most troubled times was assembled (Figures 9-1 through 9-9). Not all of the photographers involved in the project documented the social culture of this era, as some concentrated on an architecturally oriented landscape and did not include people in many of their compositions. Walker Evans is a good example of this aesthetic (Figure 9-4). However, employees such as Dorthea Lange and Ben Shahn concentrated almost exclusively on the human element in their images. Eventually, some 270,000 images were shot, constituting one of the most comprehensive photographic projects ever undertaken. Chapter 9 is devoted to a more complete explanation of this effort.

Simultaneously, other projects that both celebrated and exposed this significant time in America's visual history were initiated. A private project was completed by photojournalist Margaret Bourke–White and writer Erskine Caldwell. This collaboration concentrated exclusively on documentation of the deplorable conditions suffered by Southern tenant farm families. The photographs by Bourke–White, and the text written by Caldwell, appeared as a moderately priced paperback book in 1937 entitled *You Have Seen Their Faces*.

The FSA photographs concentrated on the positive aspects of America's small communities during the Depression. Writer Sherwood Anderson compiled a number of images to create a picture book, *Home Town*, which celebrated this attitude through use of many FSA images. However, while some American viewers considered the FSA photographs to be the truthful expression of the times, others thought them to be documents that stressed a leftist–socialist leaning and therefore disliked them as anti–American propaganda.

A Social and Political Landscape—Europe and Japan 1930 to 1940

Although the FSA was a government–funded project, and its directors had individual ideas as to its orientation, it was not originally intended to have a specific political direction. This apolitical orientation was certainly different from photographic works disseminated in Europe and the East during the same period.

During the late 1920s and 1930s, many European photographers were concerned with the economic and political evolution of their societies. Their photographs evidenced this concern. Artists distributed their images with the intention of inciting the working class into recognition of their intolerable living and working conditions. In fact, a German magazine of the time promoted the camera as a "weapon" in the leftist ideological struggle of the masses. Images with strong political content were shown widely and exhibited in areas where workers congregated, in the leftist press of the time, and even in two international exhibitions held in Prague, Czechoslovakia, during the mid–1930s.

This photographic work was inspired in part by the "documentary truth" evident in the collages of 1920s Russian constructivists Lissitzky and Rodchenko (Chapter 10). However, unlike the constructivists, many European photographers did not rely on a manipulated image to carry their message. These members of the "worker–photographer" movement were unimpressed with a sophisticated level of aesthetics. In fact, the artistic elite were scorned by these proletariat artists. They believed that only a "worker's" eye could accurately perceive a correct viewpoint of the leftist society. This utopian society was to be formed through political and economic unrest.

This political unrest took many forms. The above–mentioned leftist worker's coalition was one entity that changed the existing social structure. Other reasons for the evolution of European society during the 1920s and 1930s included the shifting European borders and populations prior to the Second World War. For example, between 1936 and 1942, thousands of Jewish people were exiled from the Soviet Union and forced to relocate in bordering countries. Roman Vishniac, a political exile of the Soviet Union, photographed this transplanted culture in the Jewish ghettos of Poland. Working with a hand camera, Vishniac was generally able to photograph unobtrusively both indoors and out. The result is an almost voyeuristic view of a culture steeped in its own religious and social traditions. When viewing the some 5,000 images, one is especially struck by the fact that Vishniac was working to the very eve of the Holocaust. His photographs document a social culture that has been almost completely annihilated.

4-12
Roman Vishniac
Granddaughter and Grandfather, Warsaw
1938, gelatin silver print
Roman Vishniac Archives, ICP

The political structure of Western Europe and Japan was certainly not as desperate, at the time, as that of Vishniac's Poland. However, images were coming forth from both locations that stressed a social culture of bleakness and despair. Bill Brandt, initially oriented toward surrealism, documented the miserable existence of workers in British mining towns. Almost concurrently, an English project called "Mass–Observation" was launched. It was designed as an absolutely objective documentation of life in the mill towns of the industrial North. In Japan, photographers were also studying the life–styles of the lower classes. Their images of beggars and street people are particularly interesting in that the lower classes had never, before the 1930s, been seen as fit for any type of photographic notice. However, by the 1940s, this project was effectively stopped by the government, presumably for political reasons.

The Modern American Social Culture—1955 to 1990

Between approximately 1942 and 1947, most of the world was preoccupied with with a major global conflict—the Second World War. As most energies—artistic, economic, and political—were expended in this conflict, social culture photographs reflected a preoccupation with images of the war and its circumstances (Chapter 11).

When the truce was signed in 1945, photographers began to visually translate their altered social culture. In Europe, where the conflict had raged, the emphasis was on rebuilding that which had been destroyed. The Allies, as the victors, were grieved by the human tragedy yet buoyed by a booming economy. This was especially true of Americans. In the United States, there began an era of pride in victory, and of reaffirmation of individual worth and economic dreams.

While studying on a Guggenheim Fellowship in 1955, Robert Frank, who was Swiss, traveled and photographed America during the apex of this era. However, he had a different vision of America than did its natives. During this journey, Frank photographed the "humor, the sadness, the EVERYTHING–ness and American–ness"[8] of the United States. The social culture Frank

4-13
Gary Winogrand
New York City
1963, gelatin silver print
©Estate of Gary Winogrand
Museum of Art, RISD

chose to photograph was not that of booming cities, or a prideful nation, but rather the mundane in the American culture: jukeboxes, drive–in movie theaters, rodeos, and luncheonette counters.

Perhaps Frank's vision was due to the fact that he was a foreigner and viewed the American culture in a detached manner. Maybe it was the fact that he was allied with the "beat" generation, forerunners of the "hippie" movement of the following decade. Whatever the reason, when his images were published as a book, titled *The Americans,* in 1959, the reaction of most viewers was immediate—and mostly unfavorable. As John Szarkowski mentions in his cover notes for a subsequent edition (1969), "It is difficult now to remember how shocking Robert Frank's book was ten years ago. The pictures took us by ambush then. We knew the America that they described, of course, but we knew it as one knows the background hum in a record player, not as a fact to recognize and confront."[9] (Note: although, unfortunately, Robert Frank's work cannot be featured in this text, his book, *The Americans* has been re–issued.)

Robert Frank worked with a 35–mm camera. He moved quickly and unobtrusively with a "snapshooters" ethic. Standard aesthetics were disregarded. The photographs were grainy; often the subject was tilted or in the midst of an action. His disregard for the standard compositional ethic and print quality, in conjunction with his choice of subject matter, led to a new compositional form in photography. It came to be described as the "snapshot aesthetic." Adherents to this new compositional form included such photographers as Garry Winogrand (Figure 4-13), who worked during the 1960s–1980s. His work parallels Frank's in his methodology of shooting as well as in the type of social culture he wished to depict.

Winogrand's images take the form of true snapshots. As mentioned earlier in this book, the term "snapshot" originally referred to gunslingers who could shoot quickly from the hip. This was the methodology employed by Frank and later by Winogrand. Winogrand captured seemingly inconsequential moments that symbolize his vision of society. His work in zoos and cemeteries shows a disembodied culture—albeit one with a certain humor.

4-14
Danny Lyon
The Line, Ferguson Unit, Texas
ca. 1968, gelatin silver print
courtesy MAGNUM
Museum of Art, RISD

Danny Lyon and Diane Arbus, two other photographers who worked in the 1960s, reveal a social culture that does not have that sidelong humor. Danny Lyon's work revolved around "outlaw" cultures. His first major body of work, *The Bikeriders,* depicted the Chicago Outlaw Motorcycle Club. As with many social culture photographers, Lyon attempted to show a particular portion of America's evolving social culture and to place it in a particular, timely, context. As he states in the foreword to his book, *The Bikeriders*, "If *The Wild One* [a 1950s movie] were filmed today, Marlon Brando and the Black Rebel Motorcycle Club would all have to wear helmets."[10] In his subsequent portfolio, *Conversations with the Dead* (1969), Lyon turned his camera on a true outlaw subculture. For two years he photographed within the Texas State prison system. The work was first printed in a prison print shop under the title *Born to Lose*. When it was republished for larger circulation, Lyon wrote a foreword verbalizing his commitment to the social culture ideal, "This work was from the beginning an effort to somehow emotionally convey the spirit of imprisonment shared by 250,000 men in the United States....In Texas it is very popular to speak of prison life back in the forties and back in the fifties. Well, someday we'll all say back in 1968 and that's what this is all about."[11]

Diane Arbus, on the other hand, did not have as clear a mission as Lyon in capturing a cultural American era. Although she stated that she wished to capture, "the ceremonies of our present,"[12] her unsettling images of modern American life in the late 1960s went beyond temporal ceremony. Although Arbus photographed suburban interiors and leisure activities, some of her most memorable images are of the dark subcultures of dwarfs and transvestites.

In depicting these subcultures, her work is reminiscent of Weegee's and Brassaï's. All three artists produced images that shocked their audience. Their intent was deliberate and successful. They would portray their subjects as specific individuals who represented a portion of a larger society. As Arbus said, "...a photograph has to be specific....There are an awful lot of people in the world and it's going to be terribly hard to photograph all of them, so if I photograph some kind of generalized human being, everybody'll recognize it."[13] This ideal of the specific subject symbolizing a more general population is a visual, emotional tool used by all social culture photographers.

In the 1970s and 1980s, young American photographers continued to embrace that specific–general ethic. Such image makers as Bruce Davidson (see Figure 6-13), Eugene Richards (Figure 7-10), and Mary Ellen Mark (Figure 7-9) shot portfolios of images that were directly designed to translate an evolving American social culture. Each of these contemporary artists chooses a "theme" or topic and works in that milieu until a statement or portfolio is fully explored.

A Global Social Culture—1950 to 1990

As a result of technological advances in communication, the world, since 1950, has become a "smaller" place. Visual and aural messages can be flashed anywhere in the world within a matter of seconds. Although most of these transmissions are functional rather than aesthetic, the technology affects our social culture. South American and Pan-American photographers have focused primarily on the rapid changes in their traditional societies wrought by economic and technological advances. Similarly, Czech photographer Josef Koudelka

4-15
Diane Arbus
Seated Man in a Bra and Stockings, NYC
1967, gelatin silver print
©estate of Diane Arbus 1972
courtesy Robert Miller Gallery, New York

portrays the changing life–style of the European Gypsy. Political concerns over the South African policy of apartheid and its effects are the basis of Peter Magubane's work. In Japan, Shomei Tomatsu, among others, is involved in documenting a traditional culture on the eve of obliteration through evolution. Like Roman Vishniac, who photographed the vanishing Jewish culture in Poland prior to the Holocaust, or Curtis whose subjects were Native Americans, Tomatsu understands that the traditional culture of Japan cannot continue to exist unaltered in this radically changing world.

As the world continues to "shrink" through technology and communication, we find ourselves a part of a melding global culture. Ideologies and national borders have become simple political boundaries, rather than obstacles to communication. The concept of the truly "foreign" has evolved. In the late 1800s, remote cultures were fascinating because of their unfamiliarity. Now, 100 years later, we possess an evolving global consciousness. It is the obligation of the social culture photographer to continue the documention of this ever–changing global society.

The facts of our homes and times... are the unique contemporary field of photography. It is for him to fix and to show the whole aspect of our society, the sober portrait of its stratifcations, their background and embattled contrasts. The facts sing for themselves.[14]
—Lincoln Kirstein
1938

5-1
Kenneth Josephson
Drottningholm, Sweden
1967, gelatin silver print
courtesy Rhona Hoffman Gallery
Museum of Art, RISD

5

Landscape and Nature

The path my feet took was lined with images, whole gardens of pictures. With exposures I picked bouquets, each more vivid than the previous.[1]
—Minor White

Scenic views—or landscapes—have been of concern to artists since scribes first set tools to writing surfaces. However, artists have not always treated the landscape as primary subject matter. Prior to the fifteenth century, most landscapes were generated simply as backgrounds for delineation of historic, political, or religious events. During ancient times, the landscape was devised (or occasionally completely invented) as a complementary setting for action, not as a singular, well-respected subject for artistic concern.

During the 1600s and 1700s, the ideal of visual space was redefined (notably through the use of the cameras obscura and lucida), but it was not until the Romantic Era (starting around 1770) that many Classical and rigid artistic rules were relaxed. This allowed for less formal renderings of natural scenes. The delineation of a natural site had evolved from merely a "setting" for action to an entity to be enjoyed for its own merits. The change signaled the birth of the "modern" concept of landscape.

Talbot, Science, and Documentation—1834 to 1844

Prior to the 1830s, generations of artists and scientists used the cameras obscura and lucida as tools to assist in the production of precision imagery. After setting up their contraptions, these practitioners would carefully trace the world as it was reflected through their lenses or prisms. Although the resulting images were fairly precise, they were still somewhat scientifically suspect because a human's hand was involved in the image tracing. This problem was solved in 1839 when William Henry Fox Talbot published his treatise, *Some Account of the Art of Photogenic Drawing—The Process by Which Natural Objects May Be Made to Delineate Themselves without the Aid of the Artist's Pencil.* The title was as intriguing as it was true. In this publication, Talbot described years' worth of the painstaking experimentation that ultimately allowed scientists and artists to generate more exacting representations of objects than an artist's pencil could ever provide.

Talbot's photogenic drawings were the forerunner to the calotype process, which he also introduced in 1839. He soon realized that his quest for recognition was to be overshadowed by Daguerre's equally revolutionary process. However,

5-2
Timothy O'Sullivan
Soda Lake, Carson Desert, near Ragtown Nevada
1867, albumen print
IMP/GEH

he also recognized that at least one aspect of his process surpassed that of Daguerre's: his calotypes were reproducible—daguerreotypes were not. To accentuate this unique aspect of the calotype, Talbot decided to produce a book. It was to be published in parts and would include both an illuminating text (concerning thoughts about his discoveries) and actual photographs. He titled it *The Pencil of Nature*. The first part was published in 1844.

The Pencil of Nature was illustrated with pictures of still objects in predominently natural settings. This is not to say that Talbot did not use his calotype process to photograph people or that all daguerreotypes were simply a means of capturing facial features. As early as 1845, French physicians were using the daguerreotype in conjunction with the microscope to delineate biological specimens. By 1851, daguerreotypes were being taken of moonscapes through the use of a 23–foot observatory telescope.

Landscapes, Monuments, and Topographics—1850 to 1860

The scientific world was particularly thrilled with early photographic processes. Natural minutiae could be recorded at one time and studied at another. Although unsentimental (and certainly at odds with the hysterical "portrait mania" that infected the world when daguerreotypes were introduced), scientists' view of the benefits of photography was also reflected by many philosophers and intellectuals. They believed that an impersonal view of the natural world through photographic representation was the key to social progress. In some instances, they were correct. For example, the many views of the natural world taken by most early expeditionary explorers were unaesthetic and unsentimental. Whether they made tracings from the camera obscura, or used the calotype or daguerreotype, the explorers strove to disassociate themselves from romanticism. They wished to precisely transcribe the landscape they encountered. In some instances, this meant that they made "panoramic" daguerreotypes—in order to show the entire scene in the same 180–degree arc that our eyes perceive. Occasionally, this meant that the photographer took a succession of up to eight plates, which would then be developed and mounted side by side to create the desired result. Other times, the 180–degree arc would be effected by physically moving the camera during exposure while the photographic plate was moved slowly in the opposite direction. Whichever case, the intent of the photographer was to record an exacting and unsentimental view of the landscape.

Other mid–1850s landscape photographers were, for commercial reasons, also more interested in detail than aesthetics. Theirs was the mission to answer the popular demand for foreign "scenics." Extraordinary natural formations (such as America's Niagara Falls or the wilds of Australia), as well as most types of "exotic" ruins and architecture were chosen subjects for these merchant photographers. Normally financed by wealthy individuals and usually taking a previously traveled route, photographers, with their burdensome equipment, traveled the globe. Aside from extraordinary natural formations, general scenes from Italy, Egypt, Malta, North Africa, Australia, and the Near East were all subjects for the scenic commercial photographer. The images they made fed a public hungry for visual information about a world that had just begun to shrink technologically.

The photographers fed the images fabricated on these voyages to engravers for third–generation reproduction or simply printed them in quantity through the calotype method. Improvements on the basic calotype process allowed these photographers to reproduce almost endlessly their imagery while allowing the "pencil of nature" to accurately depict the smallest nuance in a scene. As mentioned, most of these explorers and photographic travelers were more interested in capturing the scene than in inherent composition. Thus, most of these travel pictures were simply records—devoid of any sense of aesthetic.

Of course, this was not always the case. The Reverend Calvert Jones, a protegé of Talbot's, photographed landscapes and monuments in Italy and Malta. The images he sent back to Talbot were quite tasteful, aesthetic, and daring in their use of composition. However, it must also be noted that Jones had been trained as a classical painter. Francis Frith, avowedly commercially oriented, was another photographer who wanted to bring strong principles of art to his photographic frames. He wrote long treatises concerning compositional problems and at one point in 1859 presented a paper for publication in *The Art Journal* which he titled "The Art of Photography."

Frith traveled the world, photographing what was termed "topographical" images. A literal definition of "topographic photographs" would be straightforward images that highlighted the landscape and physical characteristics of particular places. Frith was a master of the genre and devoted much of his adult life to traveling, photographing, and selling his work. He was also an astute businessman and understood the insatiable and popular demand for "views" of exotic places. Toward this end, upon his return from a trip to Egypt in 1860, Frith opened the largest printing factory of the time. It, of course, specialized in the mass reproductions of Frith's own foreign landscapes.

Selling negatives of scenic vistas that were then printed in large factories was not the only method by which landscape photographers of the 1850s and 1860s were able to earn an income. Photographic campaigns and commissions were forming in most countries of the civilized world. Napoleon III, leader of the French nation, regarded photography as a godsend that would assist him in expanding his domestic and foreign programs. He was instrumental in hiring five photographers to record the features of medieval buildings in France. In England, although photographic programs were rarely directly subsidized by the Royal family, Queen Victoria demonstrated her support of the medium by purchasing a respectable number of photographic images and views.

Other, private concerns commissioned photographers to make views of specific areas. The impressively wealthy Baron de Rothschild hired Edouard–Denis Baldus to photograph both his new railroad line and the impressive natural scenery that the rail lines made more accessible to the everyday traveler. Countless private publishers hired individual cameramen to travel to the far corners of the world. The photographers would endure any amount of hardship to bring back views of the foreign and exotic, which the publisher would then contact print and sell either in shops or through subscription.

Burdens and Improvements:
The Technical Process—1850s

The calotype process was clearly favored over that of the daguerreotype by early landscape photographers. The reasons were twofold. First, the daguerreotype required glass plates that were heavy and fragile, while the calotype relied on a paper support. Second, the calotype was reproducible, while the daguerreotype was not. The one drawback of the calotype—blurred detail due to the paper fiber support—was overcome during the early 1850s by Louis Blanquart–Evrard and Gustav Le Gray. These two men discovered that waxing the paper prior to exposure would obscure the texture of the paper support. This technique resulted in better definition and more tonal sensitivity. Thus, the calotype was thrust into a position of favor and prominence with contemporary landscape photographers—a position it held until it was eventually supplanted by the collodion process.

Collodion answered the need for even better definition in the photographic negative and print. With its arrival, sharper and more predictable results were available in considerably less time than with the calotype. However, the collodion process had one major drawback for landscape photographers working in the 1850s and 1860s. The process required sensitizing, exposure, and development of a glass plate while the plate was still damp. Because of this, photographers carried immense amounts of equipment into the field for photographic purposes. One observer noted during an expedition, "The camera in its strong box was a heavy load to carry up the rocks, but it was nothing to the chemical and plate–holder box, which in turn was featherweight compared to the imitation hand organ which served for a darkroom."[2]

Two other technical problems that plagued early landscape photographers concerned the plate's inability to record all spectral hues accurately. Tonal shifts occurred because emulsions of the time recorded certain spectral light waves differently than others.

Another problem concerned including the sky in a landscape. The sky, while being recorded, imparted a tremendous amount of illumination—far

5-3
Francis Frith
Entrance to the Great Temple, Luxor
from *Egypt and Palestine*
1857, albumen print
MOMA

5-4
Peter Henry Emerson
The Shoot, Amwell Magna Fishery
1888, photogravure
Museum of Art, RISD

5-5
Peter Henry Emerson
from *Marsh Leaves*
1895, photo–etching
Museum of Art, RISD

more than that of the land. Therefore, if a cameraman exposed for the sky and clouds, the earthbound portion of the frame would be underexposed and without detail. Conversely, if the photographer gave his plate proper exposure to record the landscape, the sky would be featureless and overexposed in the resultant print. Photographers who wished to have a detail–filled frame had only one option. They shot two plates of each landscape. One short exposure was made of the sky. Another, longer exposure was made of the landscape. The two frames would then be combined during the printing process, resulting in a more completely filled composition. Many aesthetically oriented photographers used this method of "artifice." Some even kept a stockpile of pre–exposed "sky shots" to print into their newer landscapes.

The American West: Fact and Vision—1860s and 1870s

During the 1860s and 1870s a number of photographers traveled to little–known territories in order to document the hitherto unexplored expanse of America referred to as "the West." Some of these image makers, notably Timothy O'Sullivan and William Henry Jackson, were primarily concerned with recording "fact." They were hired by government and private survey organizations to document the topography of the region. Although many of their images have a particular beauty, the intent while photographing was one of recording visual "facts" of western landscape. Others, such as Carleton Watkins and Eadweard Muybridge, were more aesthetically oriented. These two photographers were notably interested in capturing the transcendental beauty of the area and its natural monuments.

Timothy O'Sullivan, one of the "fact finders," began his career as a photographer in Matthew Brady's New York daguerreotype studio. In 1861, he joined Brady's Photographic Corps and spent two grueling years photographing the battles of the American Civil War. After leaving Brady's employ in 1863, he continued to photograph the horrors of armed conflict by teaming up with Alexander Gardner (Chapter 11). At the conclusion of the Civil War, Gardner published *The Photographic Sketch Book of the War.* Nearly half of the 100 images published were the works of O'Sullivan.

When the war ended, O'Sullivan reluctantly returned to commercial work, but he found the repetition of the studio particularly dull after his adventures on the battlefront. He was rescued from this torpor in 1867, when a United States geological survey team was formed to document the 40th parallel (a cartographic or mapping term meaning latitude) from California to Utah. Headed by eminent geologist Clarence King, O'Sullivan joined a surveying party of forty stalwart civilians and cavalrymen. He was twenty-seven years old.

When he left on this foray, O'Sullivan was armed with two cameras (a 9 x 12–inch box type and a stereoscope outfit), 125 glass plates, darkroom equipment, and chemicals. The group first headed eastward, across the different ranges of the Rocky Mountains. It was similar to the route that had been planned for the proposed Central Pacific Railroad. The country was incredibly majestic, but terribly rugged. The trials of carrying that amount of equipment was daunting. With the four mules assigned to carry his equipment and pull his darkroom (a converted army ambulance), he climbed backbreaking terrain. A correspondent (though not on this particular survey team) wrote home about the tribulations of photographic exploration at the time: "the horse bearing the camera fell off a cliff and landed on top of the camera, which had been tied on the outside of the pack, with a result that need not be described."[3] Other less serious circumstances abounded. O'Sullivan remarked to a correspondent who accompanied the 40th parallel expedition that while "It was a pretty location to work in...the only drawback was an unlimited number of the most voracious and particularly poisonous mosquitoes that we met....Add to this the entire impossibility to save one's precious body fluids from frequent attacks of that most enervating of all fevers, the 'mountain ail'...Which of the two should be considered as the more unbearable is impossible to state."[4]

During this expedition, O'Sullivan photographed both upper and lower topographics. At one point, he photographed the Comstock lode mines—several hundred feet below the ground—using dangerous and unpredictable magnesium flares. His experiences may have been nearly unbearable, but O'Sullivan took to survey work with such zeal, that soon after completing his work on the 40th parallel he ventured to Panama with the Darien expedition. One aspect of this 1870 mission included locating the most practical place for a ship canal across the isthmus of Panama (then called Darien). Subsequently, in 1871, he joined First Lieutenant Wheeler's Engineer Corps Geological and Geographical Surveys and Explorations West of the 100th Meridian. Wheeler gave O'Sullivan a boat (which he promptly christened *The Picture*), a crew, and instructions to independently photograph the terrain surrounding the Colorado River.

This assignment must have gratified O'Sullivan tremendously, as he was finally afforded photographic freedom. Remember, he had left his employer (Brady) during the Civil War due to Brady's overbearing attitude. He had then been closely aligned to surveys that had specific scientific and geologic aims. For example, Clarence King (leader of the 40th parallel expedition) had very firmly held beliefs on the subject of geology and doubtless had many instructions regarding O'Sullivan's photographic methodology. During this last survey with Wheeler, O'Sullivan was free to photograph at his own discretion. It has been said that the images that O'Sullivan made during his tenure with Wheeler were some of his finest. Unfortunately, nearly all of the negatives were ruined during transport back down the Colorado River, then to San Francisco, and Washington.

In 1873, O'Sullivan joined Wheeler, again, to photograph the ecology of the American Southwest. In 1880, Clarence King—who had been promoted to Director of the U.S. Geological Survey—again employed O'Sullivan, although now his job was based in Washington. Although O'Sullivan was appointed chief photographer to the U.S. Treasury, his tenure was short lived. He had battled tuberculosis for a number of years prior to his appointment, and he died months after he was hired.

5-6
William Henry Jackson
Yellowstone Park—Diana's Bath with Figure of Thomas Moran
ca. 1871, albumen print
Museum of Art, RISD

William Henry Jackson, another landscape documentarian of the American West, started his career as a "colorist" in a Vermont portrait studio. Restless for the romance that the West was said to provide, Jackson left New England to become a cowboy. He traveled extensively, but when he reached the age of 24, he decided to settle down in Omaha, Nebraska. He opened a photographic studio with his brother in 1867. However, the call of the outdoors proved too strong and he ultimately left the studio in the care of his brother. Jackson struck out—without official or governmental backing—to photograph a portion of the newly completed Transcontinental Railroad, from Promontory to Salt Lake City, Utah. The resulting images were so impressive that he was soon hired by geologist Francis Hayden to accompany his group on their 1871 journey into the Rocky Mountains, the Grand Canyon, and the Yellowstone area.

Jackson spent eight years working with Hayden. His work in Yellowstone was particularly notable. He photographed its grandness and the oddity of its natural features (notably the geysers and hot springs). The images seemed almost unbelievable when made available to the general public through special subscriptions and sale in specialty shops.

The merchandising of Western vistas allowed photographers to earn additional money while on survey teams. Prior to setting out on the journey, some photographers made arrangements with their survey leaders or sponsors regarding future distribution and sales of views produced while on expedition. In many cases, when the survey concluded, the imagery was made available to the survey leader and/or the funding organization. However, any number of negatives were also retained by the photographer. This was the case with Jackson. In his later years, when he was not actively engaged in survey work, he established a very successful publishing enterprise specializing in Western images.

When Jackson's photographs became available to the public, they intensified an existing interest in the natural beauty of the West. His images of spouting geysers seemed almost unearthly, but quite worthy of belief. People who viewed these photographs were entranced by the sheer grandeur of the land. It was a beauty they wished to see preserved for themselves and their children. It is, therefore, not surprising that when Jackson printed the album titled *Yellowstone Scenic Wonders* and submitted it to the American government, action was soon taken to protect and preserve that beautiful territory. A bill was introduced in Congress to forever set aside this area as a national park. On March 1, 1872, with the wholehearted approval of President Grant, and without a single dissenting congressional vote, the bill was passed into law. Jackson was justifiably proud of his contribution toward saving this wilderness. In his autobiography, *Time Exposure,* he expressed his belief that his photographs had "helped to do a fine piece of work."[5]

William Henry Jackson, like most survey photographers, took a variety of cameras on expeditions. This was partly as insurance against accidents (don't forget the falling horse), but also because different negative sizes and formats were desirable. However, Jackson astounded even the most adventurous when, in 1875, he loaded up a 20 x 24–inch camera for an excusion into the Rocky Mountains. As such cameras are not normally in use today, it is important to note that the 20"x 24" glass negative is the size of a contemporary wall poster. As the negative was so substantial, no enlargement (thus no loss of detail) was necessary. Jackson used this enormous camera because each iota of detail was essential in order to render the Western wilderness as precisely as he desired. This self–imposed requirement for representational precision led Jackson to place human figures in most of his landscapes. Asking another member of his team to be an element in the frame was not due to overt fondness for the individual. It was, again, part of his quest for exactitude. He simply wanted to use the human form as a measuring rod to provide the viewer a better comprehension of the relative sizes of the extraordinary Western natural formations.

Jackson was not the only, or the first, to use a formidably large camera in the wilderness. Both Carleton Watkins and Eadweard Muybridge had already extensively photographed the Yosemite Valley in northern California using cameras of almost equal size.

Watkins began his photographic career in a daguerreotype studio, the owner of which specialized in scenic views of the world. In 1863, at the age of 38, he began a series of landscapes in northern California that were to make him famous. His closest competitor in this venue, Eadweard Muybridge (also known in photographic history for his stop action photographs, Chapter 3), also used a large format during the 1860s and 1870s to capture scenes of Southern and Western wilderness.

Although all four photographers, O'Sullivan, Jackson, Watkins, and Muybridge, were attracted to photographing the relatively virgin territory of the American West, there was one very pointed difference in the resulting imagery. O'Sullivan and Jackson were documentarians. Watkins and Muybridge were aesthetes. Although O'Sullivan occasionally skewed his camera a couple of degrees on its tripod, presumably to create a more pleasing composition, his was essentially the search for a simple record of the wilderness. Likewise, Jackson, in search of simple and exact topographics, used both a large camera and the human figure to accentuate relative scale. The work of Watkins and Muybridge differed greatly from that of their contemporary documentarians. Watkins and Muybridge were searching to represent the mountain landscape as a symbol of transcendental idealism, or "Landscape as God."[6]

These four photographers were representational of the philosophies inherent in the mid–1800s. On one hand, this era saw the flowering of the Age of Reason, with geologists and photographers both attempting to represent untarnished fact as it related to natural form. Concurrently, a very strong segment of the art world was more concerned with a dramatic presentation of existence—owing all beauty and light not to antediluvian geology, but to divine blessing.

5-7
Carleton Watkins
Cape Horn, Columbia River
1867, albumen print
Museum of Art, RISD

Landscape: Art, Science, and "Truth"—1840 to 1900

Divergent attitudes in landscape as well as other photographic subject matter existed since the time of Daguerre and Talbot. During the 1840s, precision in detail was lauded. The art world of the 1850s was oriented more toward the artistic "effect" of landscape (and other subject matter). For example, in 1853, Sir William Newton, speaking before the Photographic Society of London, suggested that "the object is better obtained by the whole subject being a little *out of focus*, thereby giving a greater breadth of effect and consequently more *suggestive* of the true character of nature."[7] However, the 1850s was also the era of the stereograph (Figures 12-3 and 12-4), which was regarded as the truest aid to visual photographic fact in landscape.

The 1860s and 1870s hosted the artistic struggle between landscape as topographic fact (O'Sullivan and Jackson) and landscape as a transcendent ideal of God's work (Watkins and Muybridge). The 1880s brought photographers who strove to sharply render every object. However, in 1886, Peter Henry Emerson published *Naturalistic Photography*. While certainly a photographic "purist," lens precision was not basic to his photographic theories. Unlike those photographers who strove to elicit every detail within the frame, Emerson (who was, by then, a well–respected artist) strongly suggested that photographers concentrate their focus only on the central figure in the frame. This radical viewpoint emerged from his studies of the contemporary scientist Helmholtz, who theorized that the human eye focused itself sharply only in the center. Diminished acuity was all that was possible at the edge of one's vision. Thus, according to Helmholtz, accurate renditions of nature through the use of a camera was impossible.

Emerson, who had originally studied medicine, believed wholeheartedly in the Helmholtzian theory of vision, as well as the relationship between science, nature, and truth. He was, perhaps, the first to champion photography as an art in its own right (a gauntlet picked up by Alfred Stieglitz; see Chapter 8). He believed that photographers should study nature—not other art forms—in their quest for laudable results. Finally, he held firmly to the belief that photographers can, and should, have a great deal of technical control over their images. Emerson outlined all of these ideas in *Naturalistic Photography*.

However, Emerson's theories were imperiled in 1890 when two scientists, Hurter and Driffield, discovered the immutable relationship between exposure, density, and development in film emulsion. Known then as the "characteristic curve," then as "H and D" (doubtless after the discoverers), and finally transposed into Ansel Adams' "zone system," this discovery crushed Emerson's belief that artists had almost unlimited control over their imagery. In January, 1891, Emerson wrote to all editors of popular photographic magazines, begging the forgivness of all photographers who had followed his ideal of naturalistic photography. He went so far as to reprint his message the following month in a black–bordered pamphlet that he titled *The Death of Naturalistic Photography*.

However, not even Emerson could stop the notion that he had helped create—that of photography as fine art. During the 1880s, 1890s, and into the 1900s, photographers (most notably the later pictorialists) championed the belief that the photographer, not the machine or the scientific parameters of the medium, made the image. Also, although landscapes were a favorite subject matter, it was also believed that no subject should be scorned as unsuitable for photographic representation.

Integrated Landscapes—1840 to 1930

Although untouched, untrammeled vistas were a favorite subject for early photographers, some sought to record landscapes more integrated with humans and their works. These integrated landscapes sometimes entailed manipulation of existing natural forms, as when Muybridge combined an earthly landscape from one frame with a cloud formation of another. More often, however, a man–made landscape was one in which the natural topography was somehow altered through the efforts of humans. The "scenics" of monuments and ruins in Europe and the East discussed earlier in this chapter are some examples of integrated or man–made landscapes.

As previously mentioned, Timothy O'Sullivan photographed the route of the proposed transcontinental railroad. However, the vast majority of his images dealt with the topography of the landscape prior to construction. William Henry Jackson photographed the railroad itself. In the railroad photographs of the 1860s and 1870s, the ideal of photographic landscape regressed almost to the point of prelandscape art. The railroad was the "heroic act" or "historical monument;" the natural land formations that surrounded it were of secondary importance. New bridges and buildings, the push into virgin territories of the West and Southwest, the restructuring of existing urban areas—all were subjects for the photographer of the 1870s through the 1890s. Although some of these images had an artistic orientation, in many instances, the landscape was demoted to the function of backdrop for man–made structures. Thus, as the 1870s signaled the beginnings of the Industrial Era, most nonpictorial landscape images had a documentary, rather than aesthetic, quality.

Between 1898 and 1920, a notable exception to this general disregard for aesthetics in regard to the works of integrated landscape photography was Eugene Atget. While Atget certainly did not confine himself to strict urban landscapes while on his mission as "the official photographer of Paris," his images of parks and the interaction of natural forms within man–made structures deserve mention in this context.

Born in 1856, Atget decided, at the age of 42, to record his beloved Paris on film. It was a Paris of historic buildings and monuments, of boulevards and parks. His aesthetic sensibilities in melding urban architecture with trees and other natural forms stand as mute testimony to an age that evokes tremendous nostalgia.

During this period of the 1920s, the ideal of straight, integrated landscape photography—which involved landscape with an urban orientation—was also gaining acceptance in the United States. Although there were very few photographers working at this time who limited themselves to any particular subject, landscapes with a structural or urban motif were quite popular. As this was a period of tremendous growth in America, the man–made, or integrated, landscape was often represented in the work of such masters as Alfred Stieglitz, Edward Steichen, Ralph Steiner, and Charles Sheeler. Of these men, Alfred Stieglitz was perhaps the only image maker to produce a singular portfolio without any man–made intrusion. His "Equivalents" series were images of cloud formations (and later, other natural objects), first as they represented music, then as they corresponded to the emotions (not the physical works) of humans.

When Stieglitz introduced his "Equivalents" series they were met with the same negative attitude which the critics accorded his previous portfolio of portraits of his wife, Georgia O'Keeffe (Chapter 8). It appeared that the psychological abstraction of Stieglitz's vision did not suit a particular critical aesthetic for classical, vision.

Classic Landscapes—1920 to 1990

While Edward Steichen was still producing romantic landscapes, and some of the pictorialists were continuing in their quest for softly defined photographs, Stieglitz continued to champion modern artistic vision. He opened two New York galleries in the 1920s—The Intimate Gallery (1925) and An American Place (1929). During the 17 years he owned the latter, he added only two new photographers to the roster of artists he represented. They were both Americans and they were both classical landscapists. The first was Ansel Adams.

Ansel Adams had been photographing for 20 years before his first exhibition at An American Place. Born in California in 1902, he originally trained as a classical pianist. At the age of 14, he visited the Yosemite Valley. Although he was young, his first journey to the vast wilderness of Yosemite resulted in "a culmination of experience so intense as to be almost painful. Since that day in 1916, my life has been colored and modulated by the great earth gesture of the Sierra."[8] During that early trip, he photographed with a Kodak Brownie box camera.

As a result of that first revelation he experienced in Yosemite, Adams returned every year to hike and to photograph. As he had previously decided that his career lay not with a camera, but with a piano, he studied exhaustively and gave very successful concerts. However, in 1929, he was advised to rest from the musical profession and he again picked up his camera. He compared his photographic imagery, as did Stieglitz, with musical form. So, it was no surprise that when he showed his portfolio in 1933, Stieglitz declared the images to be "some of the most beautiful photographs I have ever seen in my life."[9] Adams had a solo exhibition at Stieglitz's An American Place gallery in 1936.

Ansel Adams was only one of a number of photographers committed to the ideal of representational natural form. A small number of these photographers eventually formed a group, f/64, bound by a search for beauty in "straight" photography. Founded around 1930 by Adams and six other individuals, the group paralleled the basic tenets of precisionism, a painterly term that means absolute objectivity and exactitude in the representation of form. The group's title has a technical photographic root. The aperture (opening in the lens), f/64, renders everything in precise focus. The title was perfect for that group of image makers who were practicing what they considered straight and "pure photography."

Approximately half of the founding members of f/64 were involved with photographing landscape. Adams, Imogen Cunningham, and Edward Weston all formulated unique visual statements about landscape and the natural world. Imogen Cunningham's work centered around close–ups of plant life. Cunningham's early work (1910s) was fuzzy and much akin to the pictorialist aesthetic (see Chapter 8). However, after she established contact with Edward Weston, her focus shifted to the precisionist ethic in both her plant studies and her industrial images (Figure 10-1).

Likewise, Edward Weston photographed natural form, figure studies, and industrial imagery, although he started his photographic career in 1911 as a portrait photographer. In 1915, he saw an exhibition of modern painting that stressed clean lines and sharp angles. It was an aesthetic that struck a chord in Weston. Within a few years he was photographing industrial mills, emphasizing their clean, angular lines. By 1923 he had moved to Mexico to live a simpler life. He searched for time, peace, and ultimately a very quiet form of creativity. He photographed and also transcribed many of his thoughts in journal form, which he called his *Daybooks*. When he returned to California in 1927, he again tried to support himself as a commercial portrait photographer. As such, he was only nominally successful, for his creativity was optimal when photographing nature. A turning point in his career occurred when he found a set of nautilus shells in the studio of a painter friend. He photographed these simple shells as monumental, singular objects. Soon, other natural forms appealed to his aesthetic. Green peppers, cabbage leaves, artichokes, and other mundane forms were transformed from an otherwise unremarkable objects to the level of profound subject.

Not comfortable with the pace of urban life, Weston moved further from the city to the seacoast town of Carmel, California. There he discovered Point Lobos, a windswept peninsula of land that jutted into the Pacific Ocean. It was a setting that completely altered his previous mode of work. He moved his camera from the studio and reveled in natural outdoor light. He revised his previous concept of natural form as microcosm and moved his camera further from his subject. Dune formations, cypress tree roots, and pelican wings were a few of the many subjects represented precisely by Weston's large–format camera. In 1937, Weston was awarded a Guggenheim Fellowship, which finally allowed him to abandon the portrait business and travel the West, photographing nature. Within a few years his work was recognized both nationally and internationally. Portfolios of his works and writings were published by numerous individuals and groups (among them, Ansel Adams, who published *My Camera on Point Lobos* in 1950). Major retrospectives of Weston's work were featured in Paris and London. Tragically, however, Edward Weston was stricken by Parkinson's disease. By 1948, he was forced to stop photographing. Weston then relied on his son, Brett, who assisted him in compiling and printing many retrospective portfolios.

Brett Weston's photographic work closely parallels that of his father. Brett's inclinations early in life were to follow in his father's foosteps, using photography to describe the form and

5-8
Ansel Adams
Antelope House Ruins, Canyon de Chelly National Monument, Arizona
1947, gelatin silver print
courtesy of the Ansel Adams Publishing Rights Trust

Museum of Art, RISD

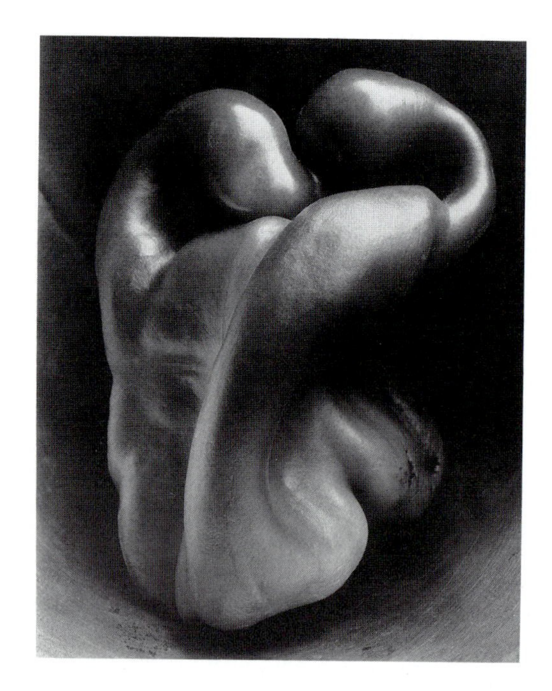

5-9
Edward Weston
Pepper #30
1930, gelatin silver print
©1981 Arizona Board of Regents, CCP

5-10
Edward Weston
Nude (Torso of Neil)
1925, platinum print
©1981 Arizona Board of Regents, CCP

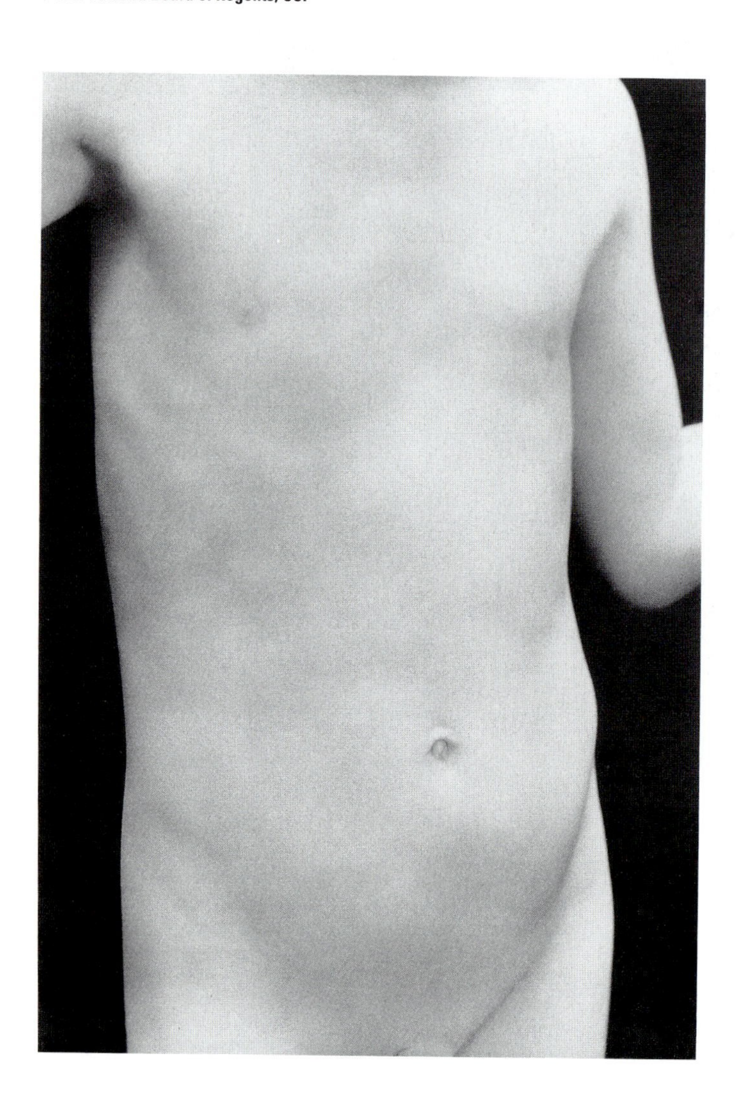

texture of nature. It is not surprising that Brett Weston's images seem to have an almost genetic kinship with those of his father. This kinship reaches from choice of subject matter to decisions concerning equipment and film. Although the majority of Edward Weston's life work involved use of black and white film, later in his career he elected to use color materials. Long eschewed as a film for "snapshot" photographers, it was used by both Edward and Brett for its ability to impart a new sense of form to photography. As Edward stated in 1947, "The prejudice many photographers have against color photography comes from not thinking of *color as form*."[10]

Personal Landscapes

Minor White, a contemporary of Ansel Adams and Brett Weston, was also consumed by a desire to represent form. White's photographic journey entailed imparting form to ideas. He was intrigued with Stieglitz's notion of the "Equivalent"— that a photograph can embody the spirit of a thought or represent a metaphor. In fact, there are many parallels between Stieglitz and White. When young, both studied the sciences. Stieglitz was immersed in mechanical engineering; White studied botany. Both experienced almost mystical revelations that drew them to photography. When White met Stieglitz, he was awestruck. When he summoned the courage to ask Stieglitz whether he could ever be considered a true photographer, Stieglitz asked him whether he had ever been in love. When White answered in the affirmative, Stieglitz affirmed that it was, indeed, possible for White to become a photographer.

It was that sense of answering a question with a question that made both Stieglitz and White such powerful photographers and teachers. While Stieglitz never actively taught, his galleries served as institutions to which photographers arrived—to show their work, to talk to "the master," and to derive conclusions about themselves and their work. White, while teaching much of his life to support himself, also enjoyed meeting with struggling photographers in order to impart a sense of self-knowledge to the "seekers."

Both Stieglitz and White were empowered with incredible energy, which they put to use in a variety of manners. Both were actively interested in photographic publishing; Stieglitz instituted *Camera Work* and White started *Aperture*. However, perhaps most importantly, both Alfred Stieglitz and Minor White had an overwhelming commitment to life. Living life properly was of primary importance. As was written in 1978, White knew that "there were more important matters in his life than making pictures, and more important goals in photography than making good ones."[11] However, both Stieglitz and White used photography as a tool to attain—and explain—personal life experience.

Most of White's images were of natural forms, although like Stieglitz and Weston, he moved easily between natural form, urban landscape, and portraiture. In his landscape imagery, like Weston, he often did not give clues to relative subject size, scale, or geographic location. Like Weston, White also found himself photographing the natural beauty in and near Point Lobos, California. Like Weston and Adams, White used a large-format camera to make his meticulously "straight" images. A contemporary and colleague of Adams, White was excited to learn of Adams' "zone system" of exposure and was an active adherent to its tenets. At one point, late in his life, he wrote an entire volume on the system and its benefit to the world of natural, unmanipulated photography.

While a strict technician, the physical properties of a photograph were only important to White in their ability to carry a symbolic message through the form of the medium. White wanted his message to be cross-cultural and available to anyone who would take the energy to understand—the symbol, the message, themselves. To this end, White (a converted Catholic) studied any number of mystical disciplines—among them the I Ching, hypnosis, astrology, Zen, and Gurdjieff's teaching on self-knowledge. He first internalized this knowledge, then attempted to impart it to the host of students who had become his lifeblood. As White remarked, "I only have students—no friends—or close ties."[12]

The list of students influenced by White's teachings is impressive. While he never insisted on specific themes or subjects, a good number of his past students are known for their work in landscape photography. Linda Conner, Paul Caponigro, Jerry Uelsmann, Carl Chiarenza, and Walter Chappell are a few among the thousands sufficiently influenced by White to go out and seek their own personal landscapes. Uelsmann, profiled extensively in Chapter 12 (Manipulation), uses a tremendous number of psychological symbols to create his personal landscapes. Linda Conner employs various nontraditional photographic techniques to enhance existent landscapes and transform them into personal statements.

5-11
Minor White
Moon and Wall Encrustations
1964, gelatin silver print
©1982 Trustees of Princeton University, all rights reserved
The Minor White Archive, Princeton University

One of White's most illustrious protégés is photographer Paul Caponigro. Caponigro is also perhaps the closest to White in his sense of mystic vision. Although Caponigro is quoted as saying, "I didn't want anything to do with all that bullshit. He [White] was dressing me up in all this masturbatory mysticism, relating my work to the Gurdjieff ideals...,"[13] the same man stated that photography "placed within a special context, can soar above the intellect and touch subtle reality in a unique way."[14] Caponigro's sensibilities were obviously touched deeply by his experience with White.

Other photographers in the mileu of unmanipulated personal landscape photography rely less on natural objects as symbols, and more on the simple composition of the frame. Aaron Siskind, for example, transforms everyday natural subjects into autonomous abstracts. His frames are tight and they emphasize the linear aspects of nature. Early in his career, Siskind was tremendously involved in the abstract expressionist movement of painting. This experience is vividly evident in the formal structure of his photographic imagery. Like Frederick Sommer's images, Siskind's photographs have little sense of physical depth. Both artists seem to prefer the depth of the image to come from the viewer's interpretation of the subject photographed.

Both Sommer and Siskind taught at Chicago's Institute of Design. Another professor of photography at the Institute of Design, Harry Callahan, was also interested in personal landscape. Callahan's landscapes, however, encompassed a broader spectrum of subject matter. He would photograph a single blade of grass as a monumental object. Later, he would pose his wife, Eleanor, in verdant natural surroundings. For Callahan, a consistent subject matter or personal "style" was not as important as the fact of photographing.

Among the more contemporary personal landscape photographers, a number studied with Callahan and/or Siskind. Linda Conner worked with both White and Callahan. Joseph Jachna studied with Callahan and Siskind. The influence of both of these instructors is evident when viewing Jachna's images of his own hand, holding variously configured mirrors. In setting up his personal landscapes, he manipulates our vision spatially, as does Siskind with his high–key images of rock formations.

Kenneth Josephson also studied at the Institute of Design, and like Jachna, introduces a portion of his own physiology into many frames (Figure 5-1). However, unlike Jachna, Josephson's images are more literal in orientation. Josephson exhibits a certain tongue–in–cheek humor when presenting his photographic subject. The same can be said for photographer John Pfahl, who introduces very particular man–made objects into an otherwise classic landscape. Pfahl has been considered conceptual in his approach to landscape. However, unlike some conceptual works of art, Pfahl's statements are visually and emotionally understandable to a majority of his viewers.

5-12
John Pfahl
Wave, Lave, Lace
1978, (original in color)
Visual Studies Workshop Gallery

The Landscape of American Society—1970 to 1990

Throughout the last 150 years of photography, various artists have attempted to portray the landscape as it related to society as a whole. Indeed, one of the most important functions of any art form is its ability to relate to societal influences. Eugene Atget was compelled to photograph a landscape as it related to the twilight of a particular Parisian society. Walker Evans photographed for the Farm Security Administration in the 1930s (see Chapter 9) and in doing so translated a particular era of the American landscape. Likewise, Dorthea Lange and Russell Lee, when working for the FSA, occasionally turned their cameras from human subjects to capture landscapes that were indicative of the times in which they worked.

American societal landscape photographers of the late 1900s are similarly interested in photographing subject matter that symbolizes, to them, the era in which they work. It is an impersonal epoch of speed and mechanization. Houses are hermetically sealed against the weather. Humans are rarely represented, although the images are full of their works. Lewis Baltz, Joe Deal, Robert Adams, and Lee Friedlander produce imagery that shows the landscape of a particular, sterile reality. It is a landscape devoid of emotion. It is not a landscape of hope.

The Landscape of Space

In the 1960s, technology enabled humans to venture off this planet of earthly landscape and into the heavens. It ushered in a new visual vocabulary concerning landscape photography. Rivers become abstract forms in the frame. Natural forms can be studied, and photographed, in radically different fashions. The images taken in space also form a new source of aesthetic appeal.

In addition, aerial or space photography offers methods of preserving this earthly landscape. Such twentieth century technology as infrared film assists scientists in urban planning. Such nineteenth century technology as stereoscopic photography helps conservationists control forest exploitation. Through twenty-first century space technology, our twentieth century landscape can not only survive, but thrive.

The nearer we get to Nature the sweeter will be our lives, and never shall we attain to the true secret of happiness until we identify ourselves as a part of Nature.[15]
—Peter Henry Emerson

5-13
Joe Deal
Untitled view, Albuquerque, N.M.
1975, gelatin silver print
IMP/GEH

6-1
Southworth and Hawes
Unknown Lady
n.d., medallion daguerreotype
gift of Edward Southworth Hawes
in memory of his father,
Josiah Johnson Hawes, courtesy,
Museum of Fine Arts, Boston

6

Portraiture

While we give it [the photographic portrait] credit only for depicting the merest surface, it actually brings out the secret character with a truth no painter would ever venture upon, even could he detect it.
—the literary character Holgrave in *The House of Seven Gables* by Nathaniel Hawthorne

Portraiture Prior to 1840

The desire to depict human form has been of artistic and historic concern since the age of cave paintings. In the prehistoric era, symbols representing humanity were drawn on tribal walls. Normally, these symbols were simple and represented an activity that engaged a group of figures, such as a particularly successful hunt. As specific individuals were not represented, the idea of "portraiture" was not an issue. This changed when civilization became more sophisticated. By the time Egypt represented the apex of civilization in the East (ca. 1400 B.C.), notable figures were depicted in their drawings through use of specific symbols. The pharaoh—or other important personage—was specifically rendered in these drawings as the figure who carried a scepter or wore a distinctive costume or headpiece.

As the quality of tools and surfaces improved, painters were able to render more exacting human likenesses. They no longer had to rely on symbols to denote a specific person. Although actions and events were still important subject matter, and were symbolized through depictions of weapons, tools, or background structures, the importance of a specific human likeness had taken precedence.

6-2
Representation of a prehistoric cave painting. Crude renderings found in the caves of Europe, Asia, and North America depict humans as generalized tribe members. The human forms are normally engaged in actions that support the tribe and its customs.

6-3
Representation of a cuneiform drawing.

This is not to imply that the likeness—or portrait—was an honest and exact record of the subject. It is probable that certain liberties were taken in immortalizing public figures on canvas. Idealization and a sense of timelessness were the main objectives in an artist's rendering. Therefore, in the artist's desire to please the subject (and garner repeat commissions), the painter would tend to overlook the existence of unflattering physical characteristics such as an unfortunate wart or facial wrinkles.

These early portraits required that the client possess a degree of wealth. Therefore, they were most often commissioned by a small segment of the population—mostly people of consequence in religion, politics, and commerce. Prior to the 1750s, the "common man" was very rarely represented for posterity. One reason was the cost of the painted portrait. Another was the simple two-class social structure of the time. One segment of the population was wealthy was and considered worthy of being recorded for posterity. The other was ignored. However, this social system, represented by either privilege or poverty, continued to evolve in the early 1800s and became more egalitarian. Citizens who were not endowed with power or wealth pressed for individual political and religious rights. A middle class was forming—one which wanted to develop a sense of individual self-worth. One obvious method of presenting oneself as a distinctive being was through portraiture. However, as most people did not have the resources to commission an oil or tempera rendering on canvas, miniature drawings (usually on an ivory surface) had to suffice. In a more photographic vein, the physniotrace (Figure 1-10), the silhouette (Figure 1-6), and the camera lucida (Figure 1-7) were employed to create portraits. These devices were quicker and less expensive than traditional oil paintings on canvas. Thus, they were used for inexpensive portraits of a rising middle class.

While these four inexpensive methods of portraiture were quite popular, each of these processes had the same flaw: they required a modicum of rendering skill. Unlike the wealthy, who desired an idealized representation of themselves, the masses wanted a true portrait of their features. In the latter case, the artist's hand was not valued when it came to portraiture. Therefore, in 1839, when Daguerre presented his revolutionary process that arrested reality, the general public embraced it wholeheartedly.

6-4
Southworth and Hawes
Portrait of Unknown Lady
n.d., daguerreotype
MOMA

Early Photographic Portraiture

By the 1840s, portrait studios had opened in almost every city in the Western Hemisphere. Most studios catered to the public demand for relatively quick and inexpensive likenesses. These were often taken in assembly–line fashion, with no regard for aesthetics. Lighting was used simply to speed the exposure time and was not considered for any potential artistic effect. Because of the number of studios offering photographic portraits, price wars ensued in some large cities. This, of course, lowered the cost and allowed the sitter to opt for additional exposures if the initial plate was not sufficiently flattering.

The vast majority of these studios were opened by individuals eager to make a maximum income despite possessing a minimum of artistic skill. However, there were a few serious portraitists who considered aesthetics as a primary characteristic in their work. The Scottish team of David Octavius Hill and Robert Adamson used the calotype process to render the 470 men of religion who withdrew from the Presbyterian Church to form the Free Church of Scotland. They also photographed lesser–known acquaintances. Each image they shot was carefully and artistically composed with attention to light and detail. While, during the mid–1840s, exposure times were occasionally at least 40 seconds long, Hill and Adamson managed to photograph their subjects in seemingly casual poses, although most assuredly the subjects were somehow braced against movement during exposure.

The American team of Albert Sands Southworth and Josiah Johnson Hawes also recognized the artistic potential in the new photographic processes. Both Southworth and Hawes had seen, in 1840, one of the first public exhibitions of daguerreotypes in Boston, Massachusetts. Both were immediately excited by the possibilities inherent in the process. Although Southworth and Hawes originally worked with other partners, their joint commercial venture was launched by 1843. It was one that developed into a diverse and lucrative business. Once established making portraits, they taught classes about the photographic process. Additionally, Southworth and Hawes ran an extensive retail business selling daguerreotype supplies. Incidentally, Hawes married Southworth's sister, who then worked for the studio as a colorist.

The studio of Southworth and Hawes was dedicated to aesthetic ideals. They boasted that they never employed an "operator" (cameraman), instead they preferred to pose each sitter personally. Their advertisements boasted that "the superiority of our Likenesses is the result of our care in the arrangement throughout, particularly of the Light."[1] Southworth and Hawes also deviated from the norm in that, unlike most other portraitists, they did not charge for each exposure made. It was their standard procedure to expose several plates of each sitter and allow the client to choose the one he or she considered most pleasing. The images not chosen by the client were saved by the partners and eventually their heirs donated them to various photographic archives in the United States.

The desire of Southworth and Hawes to "produce in the likeness [of the sitter] the best possible character and finest expression of which that face and figure could have ever been capable"[2] placed them in the forefront of their field and soon many notable Americans were seen arriving at the gallery located on Tremont Street in Boston, Massachusetts. In addition to photographing statesmen, authors, and artists, Southworth and Hawes took the camera afield to shoot landscapes, city views, and interiors. Both were also innovators in the field of photography. Southworth developed a plate holder that was capable of multiple exposures upon the same photographic surface. Hawes built the "Grand Parlor Stereoscope" for viewing 12 pairs of daguerreotypes in succession; he pioneered the concept of three–dimensional photography as well .

Samuel F.B. Morse, inventor of the telegraph, was another nineteenth century photographer dedicated to aesthetic portrait quality. After learning the process in the early 1840s, Morse opened a daguerreotype studio in New York City. He hired Southworth's original partner, Joseph Pennell, as an assistant. After Pennell left, some historians believe that Morse taught Matthew Brady (Chapter 11) the intricacies of the process.

After opening a second studio in Washington, D.C., Brady stated his aim "to vindicate true art" through the production of better portraits. His prices were much higher than those of his competitors. However, people were willing to pay, as Brady had made a reputation for himself through photographing popular figures. He then sold these dagerreotypes of the "rich and famous" to popular magazines. These were translated into woodcuts, though the imagery was always noted as having been derived from a Brady daguerreotype.

Across the Atlantic, a number European photographers also recognized the importance of aesthetics in their portraits. Gaspard–Félix Tournachon, a Parisian known as Nadar, founded a portrait studio in 1853. It was a favorite meeting place for such established artists as painters Eugéne Delacroix and Gustave Doré, actress Sarah Bernhardt, and writer Charles Baudelaire (who had called photography "the refuge of every would–be painter."[3]) Nadar, originally trained in medicine, then as a writer, and later as a caricaturist, photographed these friends with particular directness and dignity. Some of these images were used in his famous mid–1850s lithograph—*Pantheon Nadar*—an oversized image featuring 270 caricatures of recognized European public figures (Chapter 3).

Nadar's *Pantheon Nadar* was a forerunner of the collaged cartes–de–visite, which enjoyed tremendous popularity in the 1860s and 1870s. Originally conceived as photographic calling cards (Chapter 2), cartes–de–visite were presented to the host as a personal memento of a social visit. Within a few years, however, the purpose of the carte evolved from a personal memento to a commercial enterprise.

These commercial cartes featured collaged images of notable people. They were fabricated by pasting together individual carte portraits of well–known political or society figures. Occasionally, the person constructing the image had to rephotograph painted portraits in addition to using current cartes, in order to achieve a complete col-

6-5
Nadar
Pantheon Nadar
1854, lithograph
Bibliothéque Nationale, Paris

6-6
unknown photographer
Seventy Celebrated Americans Including All the Presidents
ca. 1865, collaged carte-de-visite/ albumen print
Library Company of Philadelphia

lage. These collaged cartes were usually sold for a nominal amount and were the equivalent of our contemporary post cards.

Although the collaged carte–de–visite was a popular collectible item, it never supplanted the unmanipulated carte as a personal memento. By the 1860s, in New York City alone, over 100 studios specialized in personal portraiture.. Most studio operators relied on pedestrian clients. These people would dress carefully, arrive at the studio, stand in front of one of a variety of backdrops, hold still for the requisite number of seconds, wait for the product (whether it be an ambrotype, tintype, etc.), pay for the result, and leave. Other, more intrepid, photographers made house calls. Usually these visits were reserved for famous clients (such as royal families). However, occasionally these house calls involved the macabre business of photographing the recently deceased.

Medical epidemics were rife until well into the twentieth century. They struck young and old alike, although it was especially tragic when a youngster fell victim. The families, many of whom had not availed themselves of a portrait of the youngster while living, called the studio owner to arrange a posthumous portrait. In the 1850s and 1860s, this required that the photographer transport camera, tripod, and darkroom to the site. Some photographers, taking into consideration the immobility of their subject, relied on the collodio–bromide plate—a dry plate (more convenient) process that required lengthy exposure times. The child would usually be positioned as if sleeping, showing a blissful expression and perhaps clutching a favorite toy. However, in the 1860s, a cultural drive toward increased reality in all facets of life (and death) made some clients request that the photograph be taken while the child rested in the coffin, surrounded by flowers.

6-7
unknown photographer
Portrait of a Dead Baby
nineteenth century, tintype
Museum of Art, RISD

An odd though less morbid sense of reality in portraiture was required by the medical community in the mid–1800s. Photographic documents had been made of medical problems (burns, physical oddities, etc.) for a number of years. In these images, the camera often focused on the afflicted area of the body, not on the subject's facial likeness. However, in the 1850s, Dr. Hugh Welch Diamond began photographing the female inmates of the Surrey County Asylum in England. In a paper read to the Royal Society in 1856, Dr. Diamond stressed the importance of photography for diagnosis, identification, and treatment of these patients. Charles Darwin, the father of modern evolutionary theory and author of *The Origin of the Species by Natural Selection*, made a scientific though less medically oriented series of portraits. In 1872, he published a volume entitled *The Expression of Emotion in Man and Animals*. To illustrate this text, he used photographs of his own making, as well as photographs taken by others. A particularly interesting series in this volume used Oscar Gustave Rejlander as the subject.

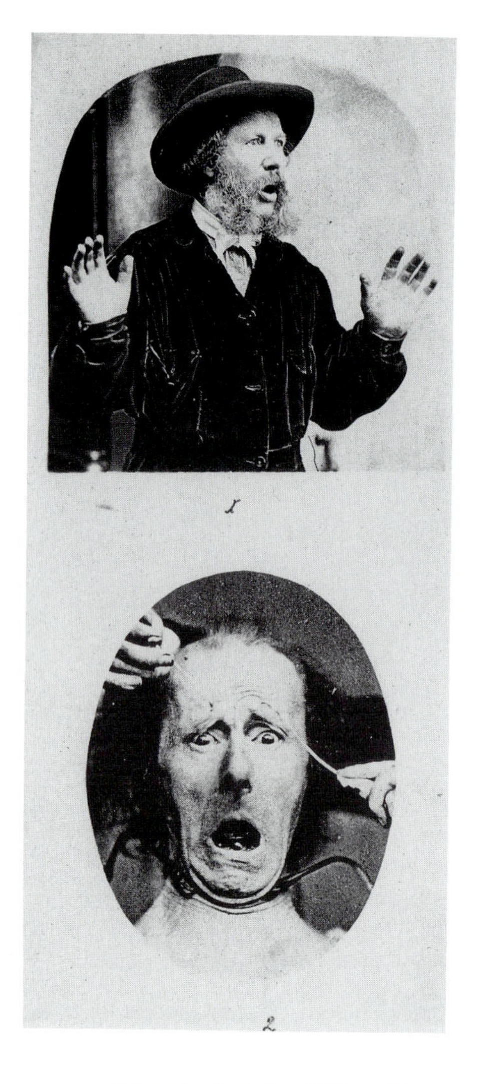

6-8
Oscar Gustave Rejlander
from *The Expression of the Emotions in Man and Animals,* by Charles Darwin
1872, heliotype
New York Public Library/ Astor, Lenox, and Tilden Foundations

Expressive Portraits—1870s to 1920s

It is not surprising that Rejlander was willing to pose for these emotive and theatrical portraits. His own personal photographic work was fraught with a sense of drama and storytelling (Figure 12-5). Although his work was savagely attacked by critics, he was one of a number of Victorian era (1840–1890) photographers who believed that personal expression should supersede reality. Another photographer of this time who revered expression over the literal in her photographic work, was Julia Margaret Cameron.

Cameron was English, though born in Calcutta, India. She grew up in comfortable circumstances, which did not change after her 1838 marriage to a high official in the British Civil Service. Her husband was a representative of the Crown in India. Cameron was soon known as a remarkable hostess and quickly made friends with many of the British gentry who visited the Camerons while on their international travels.

Upon moving back to England in 1848, Cameron maintained her correspondence with such contemporary geniuses as writer Alfred Lord Tennyson, scientist Charles Darwin, poet Henry Wadsworth Longfellow, and chemist Sir John Frederick William Herschel (who had devised the word *photography* and was godfather to the Cameron's youngest son).

Herschel had been a friend of the Camerons for many years. His interest in photography must have affected Mrs. Cameron, as when she was given her first photographic apparatus in 1864, she immediately turned her world upside down in order to accommodate her zeal for the medium. "I turned my coal house into my darkroom, and a glazed fowl–house I had given to my children became my glass house [studio]....it is my daily habit to run to [my husband] with every glass upon which a fresh glory is newly stamped....This habit of running into the dining room with my wet pictures has stained such an immense quantity of table linen with nitrate of silver, indelible stains, that I should have been banished from any less indulgent household."[4] This enthusiasm must have been a shock for her family, who had given her the apparatus merely for simple amusement.

Julia Cameron originally photographed the children and farmers around her home on the Isle of Wight. She gradually expanded her subject matter to include those famous correspondents previously mentioned. Once positioned in her studio, she held them captive. With few exceptions, this meant sitting in the hot studio for hours on end, while Cameron adjusted posture or changed focus. Tennyson, a neighbor, remarked when leaving Longfellow at Cameron's studio, "You'll have to do whatever she tells you. I'll come back soon and see what is left of you."[5]

From 1865 to 1875, Cameron decided to pose her subjects as idyllic figures from mythology or poetry. Her visual sense of the mythical was heightened by a soft, dreamlike quality. This was the result of deliberately indefinite focusing. She believed that the lack of definition and the resulting visual softness enabled her to capture "the greatness of the inner as well as the features of the outer man."[6]

This presentation of soft–focus photography was rekindled some 15 years later by Peter Henry Emerson, proponent of naturalistic photography (Chapter 10). His methods were questioned by some critics, who said that "Naturalistic focus, according to Emerson, means no focus at all, a blur, a smudge, a fog, a daub, a thing for the gods to weep over and photographers to shun."[7] However, a number of photographers disregarded the critics and actively pursued a softer definition in their images. Most notably, a number of the pictorialists and photo–secessionists (Chapter 8) actively strove to delineate a less accurate image than the camera was capable of providing. Both Gertrude Käsebier and Edward Steichen used Cameron's ideals and Emerson's naturalistic techniques during the early 1900s.

Käsebier's work bore the imprint of Cameron's influence in that it centered around an idealized figure or scene. While her subjects were never notable (as were some of Cameron's), she attempted to represent the same type of idyllic genre imagery that was reflected in Cameron's work (Figure 8-4). Käsebier's style also followed that of Cameron in its deliberate softness, although its lack of definition is the result of the paper fibers on which she printed her gum–bichromate images (Chapter 12), not of indefinite camera focus.

Edward Steichen's portraits were also very impressionistic. Like Käsebier, he often relied on the softness inherent in the gum–bichromate process to deliberately obscure detail in his imagery. Like Cameron, he often photographed famous subjects. Painter Henri Matisse, sculptor Auguste Rodin, President Theodore Roosevelt, and American financier J. Pierpoint Morgan were just a few examples of subjects idealized through Steichen's impressionistic photographic methods.

By the early 1900s, Steichen's interest in portraiture led him to open a commercial studio at 291 Fifth Avenue (Chapter 8). However, soon after the studio was established, he began working closely with Alfred Stieglitz on various noncommercial photographically oriented projects. World War I intervened in the 1910s, and Steichen served as an administrator for a branch of the United States military which dealt, in part, with aerial reconnaissance photography.

Aerial photography is strictly factual, although somewhat abstract in that it basically shows the world in two dimensions. It is evident that Steichen's experiences with these aerial abstractions had a significant impact on his later photographic style. After the war, Steichen's imagery became starker and more realistic than his earlier, softer photographs. In the early 1920s, Steichen returned to his studio in New York City. He again worked in portraiture and, by 1923, he was hired by Condé Nast publications as its chief photographer.

6-9
Julia Margaret Cameron
Madonna with Children
ca. 1866, albumen silver print
MOMA

6-10
Edward Steichen
Charlie Chaplin
1925, gelatin silver print
MOMA

Portraiture for the Printed Page

Steichen's work for Condé Nast included photographing for two of their most successful publications, *Vogue* and *Vanity Fair.* The readership of these magazines consisted mostly of affluent and middle–class women who were both fashion and celebrity conscious. Steichen's portraits of current cultural and social figures were exactly what the readership desired. Steichen's celebrities were portrayed as human, although extremely (almost unrealistically) elegant. For example, in his images of actor Charles Chaplin, composer George Gershwin, and actresses Greta Garbo and Gloria Swanson, Steichen presented the human likenesses but stressed the individual characteristics of each celebrity. When appropriate, he would use props or trademarks that were popularly associated with each personality and their respective profession. Lighting was always a critical factor, as it was in the work of his predecessor at *Vogue*, Baron Adolphe de Meyer.

It is questionable whether Adolphe de Meyer was actually born into a noble family, although de Meyer insisted that everyone call him "Baron." In fact, photographic historians to this day disagree as to his country of origin, his date of birth, and his family's circumstances. It is known that de Meyer began exhibiting his photographs in 1894, and within 15 years, he was a member of the elite Linked Ring in London.

The Linked Ring was formed by photographers who were disillusioned by the code of aesthetics favored by the Photographic Society of London (later renamed the Royal Photographic Society). "The Ring" adopted its aesthetic orientation from the artistic code promoted by P.H. Emerson and the photo–secessionists—that of a higher elevation of art for its own sake. The feeling inherent in the works of members revealed a highly poetic or impressionistic visual form. Adolphe de Meyer certainly ascribed to these aesthetic beliefs.

De Meyer originally exhibited carefully constructed still lifes and portraits. Gradually, he built a clientele of fashionable Londoners who wished to be immortalized by his soft–focus lens. Eventually, he began publishing these portraits in English magazines and then emigrated to New York in order to work for the most avant–garde fashion magazine of the day—*Vogue.*

At *Vogue,* de Meyer used his impressionistic techniques to the limit. His soft–focus lens and use of diffused light provided a tremendously romantic quality to his imagery. His sensibilities were instantly popular with *Vogue's* magazine readership. However, the public is fickle and within a few years, de Meyer's techniques were considered "old–fashioned." He left the employ of *Vogue.*

As the economic and political climate changed (American Depression, 1929–1935; World War II, 1939–1945; etc.), the tastes of magazine readership were altered. A new type of austerity and visual objectivity emerged in terms of portraiture for the printed page. Some photographers, such as Richard Avedon, Arnold Newman, and Irving Penn, responded to the changing cultural dictates of magazine subscribers. They understood that there would always be a market for mass–produced likenesses of famous cultural figures, but that the lushness inherent in the work of earlier photographers was no longer appropriate. Likewise, between the 1960s and 1990s, periodical portraitists such as Annie Liebowitz (working for *Rolling Stone* then Condé Nast) periodically change their shooting style to match the evolving sensibilities of their readership.

The Portrait as an Allegorical Document

Of course, mass–produced images involving notable subjects is only a small facet of the history of photographic portraiture. Since the daguerreotype, likenesses of the general populace have greatly outnumbered those of popular heroes. Thousands of tintypes, cartes–de–visite, and ambrotypes still exist which feature anonymous faces whose names have been lost to time. Few clues as to the subjects' personalities or livelihoods can be gleaned from these photographs. The photographers who made these images are, by and large, as anonymous as their subjects.

However, there were also a number of photographers who still are remembered for historically oriented projects that featured portraits of "the common man." Some of these photographers wanted to record social customs or cultures that would be lost over the course of time. John Thomson's set of images titled *China and Its People,* taken in 1873 to 1874, was the start of documentary portraiture. Soon after, such photographers as Adam Clark Vroman and Edward Curtis (Figure 4-6) worked to depict vanishing "native races." Through the course of this century, many individual photographers have, through portraiture, captured the life–styles of distant cultures. One common thread that binds these myriad photographers and their projects is the manner of subject treatment. In photographing their subjects, these artists are not as interested in the individual, as in the person's representation of an occupation, an ideological subtext, or a culture. Rarely do the photographs' titles correspond with the given name of the sitters. Thus, these portraits must be seen in an allegorical or symbolic context. These portraits are visual symbols of a larger society.

6-11
Baron de Meyer
Helen Lee Worthing
1920, gelatin silver print
MOMA

August Sander's work is an example of an idea that falls halfway between true portraiture and allegorical document. Sander, in 1910, began a series of documentary portraits that were meant to describe "Man in 20th Century Germany." His idea was based on the concept of a "sociological arc," which would include peasants as well as the high-born of German society. He was convinced that a true portrait of Germany would emerge through photographing its socially stratified citizens. Although Sander was a commercial portrait photographer, it is doubtful that most of his subjects paid for their likenesses or that he confined himself to professional sittings. He wanted to "speak the truth in all honesty about our age and the people of our age."[8] In 1929, he published a compendium of these portraits under the title *Antlitz der Zeit (Face of Our Time).*

Unfortunately, the changing political climate in Germany immediately following this publication forced Sander to prematurely end his photographic research. The book was banned by the Nazi regime, in part for the very reason that Sander had begun the project. Sander wished to portray both the differences and similarities of "Man in 20th Century Germany." The Nazis were interested in promoting the Aryan, or perfect, race.

6-12
August Sander
Circus People, Duren
1930, gelatin silver print
Museum of Art, RISD

Sander's images, unfortunately, exhibited too much of a composite portrait and not enough of the monotypical "blue–eyed, blonde–haired pureblood" who was purportedly destined to govern Germany's future. Unable to continue his chosen project, Sander turned his camera away from the human figure and pointed toward landscapes and industrial scenes.

Sander documented a culture and subculture through symbolic portraits. Although many of his images do carry the name of the sitter in the title, the vast majority are of people who are represented only by their occupation or location. The desire to describe culture through individual members is common among many photographers of this century. Lewis W. Hine is one—in his early child labor imagery as well as his subsequent book, *Men at Work.* Dorthea Lange is another. In these cases, the subject is symbolic of the events of an era, not an individual ego being immortalized.

Dorthea Lange began her photographic career in the 1920s as a studio portraitist. In her early commercial work, it is evident that her subjects were simply individuals paying for a portrait. However, by 1932, Lange was photographing on the streets of San Francisco rather than in the comfort of her studio. She was drawn to what she referred to as "imperfect people"—those who showed, through gesture, clothing, and features, the physical and social circumstances in which they lived. The American Depression loomed large in the economic and political times. Lange, through her compassionate portraits of individuals and families, demonstrated the Depression's adverse effects, as well as the strength of the human spirit in overcoming these difficulties (Figures 9-8 and 9-9). At the conclusion of the Depression, Lange centered her photographic interests on another set of individuals—Japanese-Americans who were unjustly interned in closed "camps" during the hysteria that evolved almost immediately after the bombing of Pearl Harbor. As with her photographs of Depression–era people, the Japanese–American portfolio demonstrated her subjects as allegorical symbols first, as individuals second.

This aesthetic–social–allegorical idea was carried into the 1960s by Bruce Davidson, among others, who photographed the Civil Rights struggle

6-13
Bruce Davidson
Untitled
ca. 1967, gelatin silver print
MOMA, courtesy MAGNUM

of black Americans. He later confronted social issues through photographic portraiture by shooting a series on the inhabitants of a run–down section of New York City. His artistic and humanistic concerns paralleled those of Lange. His subjects were certainly of the lower economic strata. In fact, East 100th Street was considered one of America's most deplorable slums. Davidson, a white photographer in a completely ethnic neighborhood, was highly visible. Those who lived in the neighborhood occasionally uttered such hostile remarks as, "You may call this a ghetto. I call it my home."[9]

However, Davidson's purpose was not to show his subjects as objects of scorn. Rather, his images resemble Lange's in their forthright method of presenting dignity in adverse situations. Again, as with Lange, his images were rarely titled with the name of the subject. The subjects were symbolic of a wider sociological context.

There have been many photographers in the 1900s who depicted their human subjects as allegories of a larger society. Some were completely involved in the human condition. Others, such as E.J. Bellocq (Figure 4-7) and J.H. Lartigue (Figure 4-9) made early 1900s portraits in a cultural context, although it is doubtful that they did so for any historical perspective. In the 1960s, such photographers as Bill Owens, in his book *Suburbia* took a more lighthearted view of his subjects in relation to their culture. Still others, notably Diane Arbus involved themselves with portraits in which the photographer's psyche is evidenced as strongly as that of the subject.

6-14
Bill Owens
from *"Suburbia"*
ca. 1980, gelatin silver print
collection of the artist

The portrait can deal, therefore, with any number of factors in a number of different priority structures. With Arbus, cultural subject (as a symbol) comes first; the photographer's psyche, second; and the physical attributes of the "sitter," third. (Figure 4-15). Other portraitists balance these three factors differently. Occasionally, they meld into the ultimately personal portrait—the self-portrait. Also, there is a category of portraiture in which the photographer enters the frame viscerally, not visually. These are regarded as "family portraits."

6-15
Harry Callahan
Eleanor
ca. 1963, gelatin silver print
Museum of Art, RISD

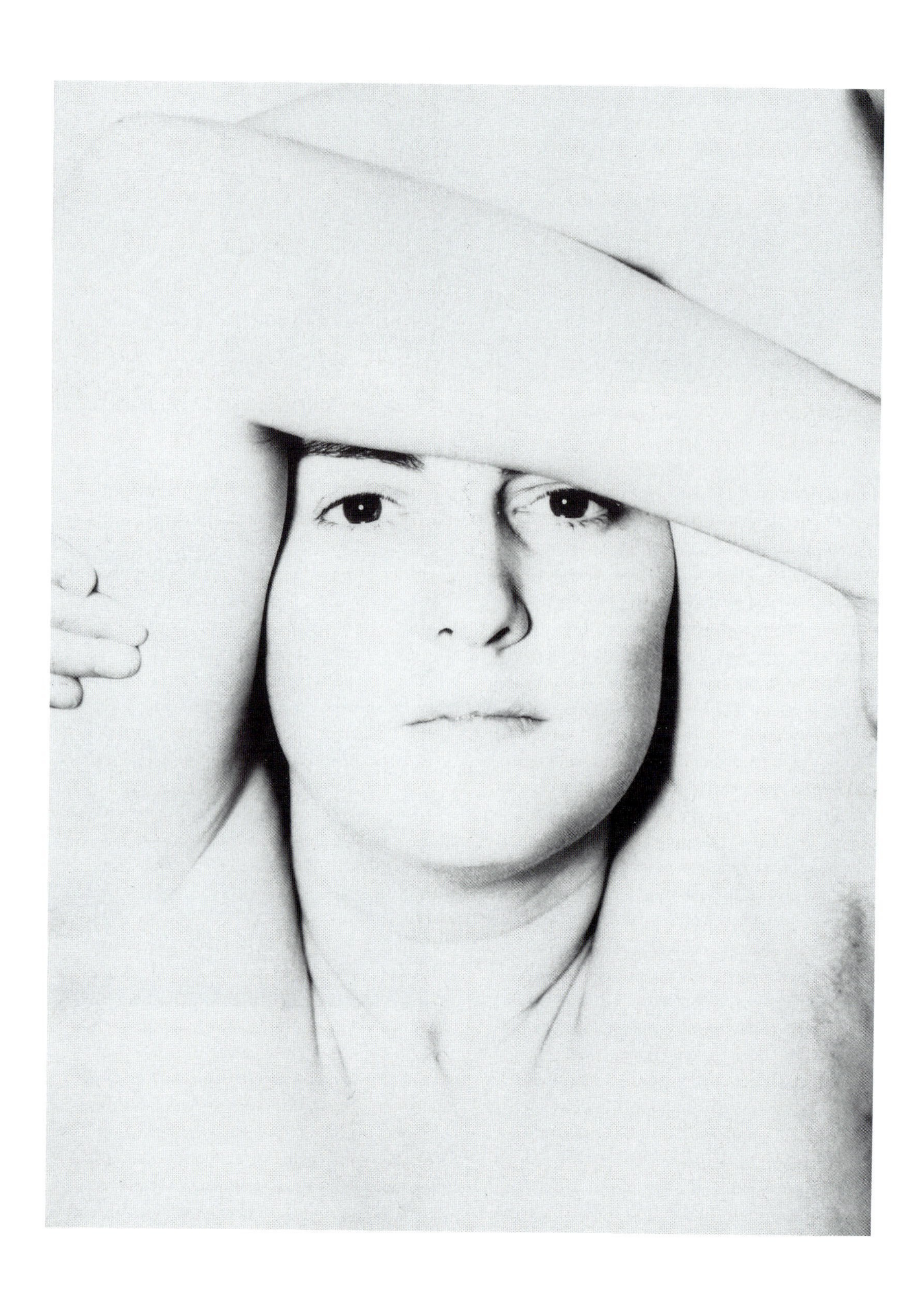

6-16
William Wegman
Dusted
1982, from Polaroid 20 x 24 color print
courtesy Holly Solomon Gallery

Family Portraits

Since the time licenses were first purchased for the right to use the processes of Talbot and Daguerre, photographers have gathered their loved ones to compose an image of the "family." Nearly 50 years later, the introduction of the Kodak #1 camera permitted almost anyone with the inclination to shoot images of their families, whether formally or informally. Millions of people have a personal archive—kept in dusty boxes—of portraits that portray a generational past. Most were taken in a casual manner, as most of the image makers were not professional photographers. However, over the past 150 years, there have been certain photographers who have used their family members as deliberate subjects. They have incorporated their loved ones into their professional work.

The listing of such photographers would be nearly endless. Certainly Julia Cameron and J.H. Lartigue (Chapter 4) must be considered primary participants in this venue. Others, working later in the 1900s, while not considered portraitists, still incorporated images of loved ones into their photographic works. Edward Weston, known primarily for transforming common objects into heroic compositions, photographed family members (most notably his sons) throughout his career (Figure 5-10). Alfred Stieglitz, founder of the photo-secession movement, considered his portraits of Georgia O'Keeffe, his wife, to be among his strongest works (Figure 8-7). Harry Callahan concentrated his camera on his wife, Eleanor, and daughter, Barbara, for an extended period. Most of these artists along with many others, worked between the 1920s and the 1950s.

During the 1960s, with the sophistication and general availability of equipment and the universal recognition of photography as a legitimate art form, younger photographers continued to experiment with the concept of photographing loved ones. Emmet Gowin, a student of Callahan's, has based a majority of his work on photographing his wife, Edith, in various situations. As with Harry Callahan, he does not shoot in a standard photographic studio. He also uses his wife as a "model" within the frame. While California photographer Judy Dater does not limit herself to family members, she, too, photographs loved ones as models in a nonstudio context.

Richard Avedon, respected for decades as a commercial photographer, shot a very personal series of images that portrayed the ravages of cancer on his father. Following the style of his previous work, he used a white studio backdrop to eliminate details—forcing the viewer to confront the devastating effects of fatal disease. Robert Mapplethorpe also uses a studio setting for his disturbing portraits of homosexual friends in a variety of unusual poses. This work provided a considerable amount of controversy in the early 1990s. William Wegman expanded the idea of family portraiture in the studio with a series on his dog, Man Ray. Wegman used his dog as a model and placed him in odd circumstances, juxtaposed with incongruous objects or dressed in outlandish outfits. On the death of Man Ray in 1982, Wegman paused in his work, then began a new series—along the same vein—of his new canine model, Fay Ray.

During the 1980s, Eileen Cowin, another contemporary photographer, concentrated on the "Family Docu-Drama Series." In this portfolio, she photographed her husband, two stepchildren, and her identical twin sister in a series of emotive tableaux. They were usually shot in her own home and were designed to combine fact and fiction. In using her twin sister, as well as those closest to herself, she incorporates the idea of "family portrait" and "self-portrait."

6-17
Cindy Sherman
Untitled Film Still
1979, gelatin silver print
courtesy Metro Pictures

Self–Portraits

Since the earliest days of photography, image makers have turned the lens toward themselves at some point in their careers. Hippolyte Bayard, in 1840, made perhaps the first effective use of self–portraiture when he constructed the image of himself as a drowned man (Figure 1-11) to protest the neglect afforded his early photographic discoveries. It would be impossible to recount all of the memorable self–portraits generated throughout the last 150 years. Certainly, when George Eastman made it possible for the general population to experiment with photography in the late 1880s, the number of photographers wanting to capture their own features increased dramatically. When "photo booths" were introduced to arcades, bus stations, and other public places in the late 1940s, anyone with 25 cents could pose before a completely automatic camera. Within minutes, the subject would retrieve four self–portraits—processed and nearly dry. It certainly did not reflect an art form, but it was accessible to a culture that desired speedy and efficient self–portraits.

During the 1970s and 1980s, such photographers as Cindy Sherman and Bruce Charlesworth used themselves in their imagery, much as Eileen Cowin used her family in her "Docu–Drama" series. Cowin's work was based on the concept of the "television docudrama." The work of these newer photographers also acknowledges our media–oriented contemporary society. Charlesworth poses himself as a character—always in peril—as if taken from a Grade–B movie script. Cindy Sherman began her self–portrait series in 1977 by presenting herself in a variety of "movie heroine" roles. Later, she featured herself in a series of self–portrait/photo illustrations based on "modern romance" magazines, "trying to make other people recognize something of themselves rather than me."[10]

Dematerialization and the Computer Portrait

While Sherman and Charlesworth photograph themselves as cultural figures (self–portraiture would seem almost a convenience in these cases), other contemporary photographers fracture images of themselves to reflect a more psychological ideal. Duane Michaels uses the concept of the picture story in his photographic series work. Joyce Neimanas (Color Plate 8) occasionally uses images of portions of her own torso in her collaged Polaroid fantasies. Lucas Samaras, also involved with Polaroid materials, has photographed himself in a fragmented vision since 1976. During the 1970s, Samaras manipulated the materials to extend a psychological vision of himself. During the 1980s, he physically extended the frame, slicing 8 x 10–inch color images of himself into long strips, which he then reconstructed into panoramas. In his work, as well as the work of those previously mentioned, the portrait becomes a medium that explores much more than the likeness of the sitter.

With the advent of the computer, we enter a period that some could consider "anti-portrait." This term describes composite or manipulated images of people fabricated through the use of new technology. This is not to imply that the idea of composite photographs is revolutionary. In the period 1877–1879, Francis Galton (a cousin of Charles Darwin) produced composite likenesses by using 10 to 12 separate negatives of different people. His project was intended to be a serious study in social science, designed to promote the theory that behavior and character were linked to physical characteristics.

The work of contemporary photographers tends to take a more lighthearted view of this idea. Robert Heinecken (Chapter 12), in such images as *CBS Docudrama: A Case Study in Finding an Appropriate Newswoman*, (Color Plate 7) superimposed the features of current television newswomen over those of contemporary anchormen. Other photographers combine facial likenesses through the use of modern technology. Nancy Burson, for example, has printed a number of computer–generated, collaged portraits that use the features of Stalin, Mussolini, Mao, Hitler, and Khomeini within a single facial likeness.

6-18
Nancy Burson
(with David Kramlich and Richard Carling)
Big Brother
(Stalin, Mussolini, Mao, Hitler, and Khomeni)
©1983, gelatin silver print
taken from monitor
courtesy Jane Baum Gallery

Although Burson's portraits are completely fictional, they must be placed within the context of portraiture. The one common bond between all portraitists in a historic context is their ability to go beyond the physical features of the model and find therein the personality of their times. A portrait can, and should, show the soul of an era—through the face of its individuals.

Miss Lydia Louisa Summerhouse Donkins informs Mrs. Cameron that she wishes to sit to her for her photograph...Should Miss Lydia Louisa Summerhouse Donkins be satisfied with her picture, Miss Lydia Louisa Summerhouse Donkins has a friend...who would *also* wish to have her likeness taken.[11]
—Note sent to Julia Margaret Cameron from Miss Lydia Louisa Summerhouse Donkins

7-1
Lewis Wickes Hine
Hine Photographing
n.d., gelatin silver print
Museum of Art, RISD

7

Photojournalism

If I could tell the story in words, I wouldn't need to lug a camera.[1]
—Lewis W. Hine
1874–1940

A photojournalist is defined as one who produces images for publication through the use of a camera. The definition differs from that of a photodocumentarian in that while both are capturing current issues, the photojournalist does so specifically for publication. Photojournalists are witnesses of specific events. They translate those events into visual imagery that is usually somewhat emotive and normally readily recognizable without a great deal of interpretation. Photojournalists must be aware of current issues, must research topics of concern, and finally, they must accept the responsibility of shaping the psyche of current culture.

Early Visual Journalism

The need for visual journalism has existed for many centuries. Thousands of years ago, "pictographs" (ancient or prehistoric drawings) were fabricated in cave dwellings. The "artists" used crude paints and sticks to immortalize such important events as battles, deaths, and successful hunting forays. A few thousand years later, ancient Egyptians made illuminated manuscripts, also to visually record important events. By 868 A.D., the Chinese were incising blocks of wood, inking them, and applying them to paper. These woodblock illustrations of important events were usually printed in conjunction with text and could take the form of scrolls or accordion–pleated stacks.

This method of supplying visual information through woodblock printing flourished in many forms for nearly a thousand years. As tools grew more sophisticated, the quality of the woodcut became finer. By 1861, publications illustrated through woodblock printing were ubiquitous (Figures I-2 and 4-2). The importance of visual illustrations is made evident by the names of the tabloids—*London Illustrated News*, or *Leslie's Illustrated Newspaper.* These publications wanted to differentiate themselves from the less visually interesting offerings of publishers who simply presented text. In 1879, *Century Magazine* announced that it was experimenting with the transfer of photographs onto wood blocks in preparation for the engraver. The announcement was met with a good deal of scorn by academic wood engravers of the day. W.J. Linton, in his volume *Wood Engraving, a Manual of Instruction*, wrote, "If you would have the best copy of a picture, do not be content with a photograph! The 'correctness' of photography is a mistake...The artist–copier translates...into black and white, so making a truer copy than the photographer's."[2] Linton was obviously afraid the new process would compromise his livelihood—much as were painters who saw the advent of photography as something to be feared (Chapter 3). Even a few publishers and editors of the day were not convinced of the necessity for photographic illustration. However, *Scribner's Monthly Magazine* answered these critics in its March 1880 issue which stated, "Mr. Linton's conservatism is apparent....The very men whose experiments Mr. Linton decries may yet prove that, even with his methods, better results are obtainable (through photography) than his school has yet produced."[3]

Although both Daguerre's and Talbot's inventions were available to the public by 1840 (Chapter 2), neither process was appropriate for mass publication of news–oriented events. Daguerre's process produced unique positive images. Because they were singular, they were of no use in the world of publication. Talbot's reproducible negative–positive process, however, seemed to be much more suitable to mass–produced publication. When Talbot patented his "polyglyphic engraving" idea in 1852, he considered its possible use within the publishing industry. Although polyglyphic engraving could produce printable steel plates from negatives and was able to record highlight and shadow areas with clarity, it had two serious flaws in relation to use for publication. First, polyglyphic engraving could only record gray middle tones through complex machinations that included double exposure. Second, they could not be printed simultaneously with printer's type. Type, at the time, was manufactured in "relief"–with raised surfaces that would catch the printer's ink. Talbot's process could not accommodate this necessary raised surface. In order to provide any sort of photographic illustration, each photograph was printed separately from the text. It was then trimmed and pasted onto the page, normally directly over the preprinted caption. In terms of labor, it was time–consuming, and therefore very costly, so it was only useful for very limited edition publications.

Thus, Talbot's process was not able to answer the immediate dictates of the publication industry to provide accurate photographic representation of current events. It was, however, helpful in that it served as the basis for other experiments centered around the nettlesome problem of presenting photographs on the same surface as type.

Photolithography (or collotype) was one of the first attempts at printing photographs and type on the same surface. Based on a 1796 discovery that ink impressions could result through the repulsion of oil and water on a stone, photolithography (later called albertype or heliotype) was in use by 1872. Although not completely compatible with type, it did permit mass–distribution of imagery. A modified version of the original photolithographic process made possible the first mass–printed photographic illustration in the March 4, 1880, edition of the *New York Graphic* newspaper.

A further advance concerning printing photographs in conjunction with type occurred through the experiments of Frederick E. Ives. Ives double–exposed a finely ruled screen in conjunction with a photographic image. The result was a relief block that could be printed with type. It signified the dawn of halftone printing.

The Photographic Halftone Process

For the first halftone image published in 1880, a simple, single–lined screen was used. These lines were detectable in the resultant photograph but the process was eventually refined sufficiently to diminish that visual flaw. Soon it was possible to produce results that closely resembled those of our current halftone process (Figure 7-3).

Basically, a halftone converts a photograph (or any other type of drawing or diagram) to a series of dots. This is done by using a camera that has been fitted with a halftone screen. The original photograph is rephotographed and the screen translates the photographic tones into "dots," which appear in varying size, depending on the tones in the original photograph. The clusters of dots usually are imperceptible unless viewed through a magnifying device. The resulting negative is then printed on a metal plate coated with sensitized liquid. When the plate is dipped into etching acid, each single dot remains on the surface of the plate, ready (after a few more steps) to be placed in conjunction with type and printed.

As mentioned, the first photographic halftone image printed in a newspaper appeared in 1880. Although adoption of this new technology was slow, within 10 years nearly all global publications were using halftones on a regular basis.

The obvious reason for the adoption of the halftone process was technical. It was simply the most efficient method of providing visual illustrations of text. Another reason that halftoning became popular is the vivid reality inherent in most

7-2
Jacob Riis
Five Cent Lodgings, Bayard Street
ca. 1889, gelatin silver print
Museum of the City of New York

7-3
Portion of an enlarged halftoned photograph.

photographic images. Prior to use of halftones, photographs were printed on wood blocks, then engraved by hand. The "artist–engraver" who cut these plates would often take liberties with the original image. If the photograph were of a burning building, the engraver might decide that, for the sake of drama, he would embellish the original with depictions of people leaping out of the windows to escape the flames. While dramatic, the public learned to accept these illustrations with a jaundiced eye. When photographic halftones were used instead of engravings, the news became more visually "real."

Photojournalist Crusades—1868 to 1924

When the public demanded "reality," most people meant that they enjoyed viewing photographic depictions of famous people and foreign places. Many members of the public were not prepared for certain social uses of the printed image. By the late 1880s, certain photojournalists were not content to provide simple, enjoyable images. Their social concerns were clear and were exemplified by two photographers in America, Jacob Riis and Lewis W. Hine.

Jacob Riis was never trained as an artist or photographer. Born in Denmark, he journeyed to America and found work as a carpenter. Eventually, he changed careers and became a reporter for the *New York Sun* newspaper. He was soon assigned to cover the Lower East Side of New York City—at that time, one of the most overcrowded, dirty, and dangerous areas of the country. Millions of immigrants had settled there in search of jobs and new lives. What they found was a trap of poverty. When Riis's initial articles describing this squalid area were published, he was accused of exaggerating the indignities suffered by the immigrants. He then turned to the camera both as a means of documentation and as vindication of his written reportage.

In 1888, the *Sun* published twelve line drawings translated from Riis's photographs under the title "Flashes from the Slums." Riis also made his photographs into slides, which he then projected through use of what he called his "magic lantern" (a forerunner of the modern slide projector). He used these slides in impassioned lectures he gave in churches and schools. Through publication and lecturing, Riis hoped to arouse public support for change in the Lower East Side. By 1890, a book of his images appeared under the title *How the Other Half Lives*. Reaction to Riis's book was immediate and forceful. Public sentiment forced Theodore Roosevelt, then president of New York's Board of Police Commissioners, to take action. Certain sections of the city were razed and parks took their places; tenements were gradually replaced by improved housing. Yet Riis was still unsatisfied. He continued to take photographs, using his camera as a weapon against indignity and the lack of social reform. His was a crusade, albeit a crusade beset with a number of technical problems; the relatively unsophisticated immigrants were undeniably nervous at the thought of being recorded by the camera. Occasionally, Riis had to bribe them with the offer of a meal. The tenements, basements, and alleys where he photographed were grimy and dark (Figure 7-2). As he needed an additional form of illumination to optimally record the conditions, Riis turned to the only existent form of portable light—flashpowder.

Dangerous and nearly uncontrollable, Blitzlichtpulver (flashpowder) was invented in Germany only a few years before Riis began his work. Flashpowder was a mixture of guncotton and magnesium powder that, when ignited, produced an instantaneous flash of brilliant light (Chapter 2). Therefore it must have been a frightening occurrence when Riis turned his camera toward the unsophisticated immigrant. As reported in the *Sun* on February 12, 1888, "a mysterious party has lately been startling the town o' nights. Somnolent policemen on the street, denizens of the dives in their dens, tramps and bummers in their so–called lodgings...have in their turn marvelled at and been frightened by the phenomenon. What they saw was 3 or 4 figures in the gloom, a ghostly tripod...the blinding flash, and then they heard the patter of retreating footsteps."[4]

Riis worked with "the ghostly tripod and blinding flash" for approximately ten years. In total, seven books of his work were published, often in several editions. His work was distinguished by powerful portrayals of the urban landscape. It was a humanitarian crusade.

Riis was not alone in his strident efforts to expose living and working conditions for the immigrant poor in New York. A fellow documentarian, Lewis W. Hine, photographically presented another aspect of life in that metropolis. Hine used a newly introduced Graflex camera, which permitted portability and exact framing, to explore the life of these "new Americans" in their struggle for social integration (Chapter 4).

Newspapers and Photographs

Our appetite for news is almost as old as the human race. In Rome, as early as 449 B.C., the Senate deposited official records of its transactions in the Temple of Ceres, where copies were made for distribution. Later, reports of gladiatorial events and gossip were added. In 60 B.C., Julius Caesar decreed that a daily paper should appear. As the printing press had not yet been invented, slaves inscribed plates that told of these daily happenings. Centuries later, Johannes Gutenberg's invention of movable type (ca. 1440) made it possible to efficiently print "news" for distribution. However, technology that would permit fast and accurate news dissemination did not exist until the late 1880s and the advent of the syndicated press and the wire service.

One of the first methods of transmitting photographs for information was used during the siege of Paris in the Franco–Prussian War of 1871. Undeveloped film was strapped to the legs of carrier pigeons that would fly out of the besieged city to a safer area. The film was then developed, and a speedy and news–worthy glimpse of the battle was available. Since that time, the concepts of speed and "news" have become synonymous and an objective of photojournalists. The Photographic Wire services served both purposes.

The idea of "pictures by wire" first surfaced in a memorandum written in 1924 by an employee of the American Telephone and Telegraph Company (AT&T). Titled "Notes on Phototelegraphy," the notice described experiments accomplished in the area of picture transmission through the use of long–distance telephone wires. Eleven years later, the Associated Press (AP) began its own network of phototelegraphy, they named it the Wirephoto. Using a network of 38 member newspapers, AP formed a 10,000–mile circuit, using wires, transmitters, and "electric eyes" to institute the world's first news photograph consortium.

Smaller Cameras, Smaller World

The Wirephoto allowed for speedy transmission of photojournalistic images. Another development, in 1924, greatly assisted the photojournalist in the areas of speed and convenience—the introduction of what was then called a new camera for "existing light" photography. The first of these, the Ernox (later called the Ermanox), was introduced in Germany. A camera that used plates was fitted with a shutter that could accommodate exposures of 1/1000 of a second and had a comparatively wide (and thus, quick) maximum lens aperture. Soon after this invention, the German Lunar camera, with similar features, was marketed. These cameras, while allowing photographs to be taken in low–light situations, were still considered a bit awkward as the film was on single plates loaded into metal film holders. The photographer was forced to stop after each shot, unload, then reload fresh film into the camera. With the introduction of the 35–mm roll–film camera—also in 1924—this situation was remedied.

The Leica, from Germany, was the first of the simple, portable, and fast cameras to be introduced that used roll film. In 1932, the Contax (also from Germany) was inaugurated. Because these cameras used a roll of film, as opposed to plates, the photographer was able to shoot continuous exposures very quickly in any given situation. These cameras also featured the fast shutter speeds and lenses that had been developed for the Ermanox. This new camera technology was instantly adopted by photojournalists around the world.

7-4
Dr. Erich Salomon
Visit of German Statesmen to Rome
1931, gelatin silver print
IMP/GEH

Photojournalism in Europe—1925 to 1934

Erich Salomon, from Berlin, was one of the first noted photojournalists to take advantage of these advances. Originally working with an Ermanox, he eventually switched to the convenience of a Leica. Salomon, in 1928, worked for the *Illustrated Press*. His particular interest was photographing European statesmen discussing business of global import. When he first requested permission to take photographs during these sessions he was refused, as the officials could not believe that he would be able to photograph indoors without a distracting flash. Determined, he sneaked into a few sessions and the resulting imagery convinced most officials that, indeed, "existing light" photography was possible through the use of these "fast" cameras.

However, some officials were not so easily persuaded. In those instances when Salomon was refused permission to photograph, he continued furtively to make images. He employed an interesting methodology to gain entry to closed sessions. He described his methods as follows: "First, arrive late. Second, dress well. Third, speak in an unintelligible tongue. If they can't understand you, they will assume you are supposed to be there."[5] Occasionally, even those ploys did not work and Salomon had to be content with moving catlike around the outside of a building, shooting through windows or curtains to catch the demeanor of the diplomats as they discussed world policy.

Felix Man was a kinsman contemporary of Salomon. Like Salomon, he was attracted to photographing with an Ermanox, but later switched to a Leica. Both he and Salomon were oriented toward recording political figures with their cameras. Both were forced to leave Germany when the Hitler regime came to power. Fortunately, unlike Salomon, who left only to be deported back to Germany (he was exterminated in a concentration camp), Man emigrated to England. He was interested in the "picture story" idea of photographing and to this end submitted work to a variety of publications.

Felix Man was born Hans Baumann. During the early portion of his career, he used both names for financial reasons. He was originally employed (as Baumann) by a publishing house that would not permit him to sell his work to other businesses. Thus, Felix Man was invented. The first photographs he showed (as Man) were to a notable photography editor, Stefan Lorant. Lorant enjoyed the spontaneity of the images and the visual flow between pictures. An English picture editor who also saw the work of Salomon and Man dubbed the imagery "candid." This word has remained in our visual vocabulary to describe quick, unposed imagery—the type of work for which both Felix Man and Erich Salomon were famous.

Salomon's "candids," Man's ideal of a "picture story," and Lorant's layout expertise ushered in a new and exciting period of design in the picture press. The hub of this period was centered in Germany from the late 1920s to the mid–1930s. By 1930, the popularity of the new design ethic was evident through the sheer numbers of magazines produced in a single press run. The combined circulation of the three most popular German picture magazines, *Berlin Illustrated Newspaper* (BIZ), the *Munich Illustrated Press* (MIP), and the *Worker's Illustrated Newspaper* (AIZ), was five million copies. It was estimated that over five miles of newsprint was used for each issue.

These magazines emphasized a collaborative effort between employees, unlike many photojournalist assignments of today. During the 1920s and 1930s, the picture editor discussed the assignment with the photographer in very vague

Münchner Illustrierte Presse

MUSSOLINI BEI DER ARBEIT

7-5
Felix Man
Layout from *Munich Illustrated Press*
ca. 1932
private collection

terms, leaving choices of coverage to the picture maker. Once the photographs were turned in to the editor, however, the collaboration stopped and the editor positioned the images using his own distinctive style. As each editor had a unique layout ideal, it was often quite easy to distinguish—without looking at the masthead—who had structured the article.

The ideal of distinctive style during this period was not confined to layout and photography. Each of these German magazines had a political stance. Of the three largest German magazines (BIZ, MIP, and AIZ), the *Worker's Illustrated Newspaper* (AIZ) was by far the most radical in terms of its political orientation. Along with its avowedly communistic photo–reportage, it featured photomontages by John Heartfield (Figure I-3) on the cover. These montages by Heartfield were designed to raise the political consciousness of even the most casual reader. During Hitler's rise to power, AIZ was blatantly anti–fascist. When Hitler eventually became the ultimate source of power in the German state, AIZ was forced to move its offices to Czechoslovakia in order to continue publication.

Of course, Germany was not the only country to produce remarkable photojournalistic magazines. In France, publications that relied on photo–reportage were also surfacing. The most popular of these was *Vu (Seen)*. It was founded in 1928 by Lucien Vogel, who stated that it should be "at once a form of expression and a means of action."[6] To this end, Vogel hired Alexander Liberman (who later worked on America's premier fashion magazine, *Vogue*) as artistic manager. Liberman's sensibilities closely matched Vogel's and he hired photographers Robert Capa, Henri Cartier–Bresson, and André Kertész, among others, to promote photo–reportage as "the form of expression."

Life Magazine

In America, magazines that relied heavily on photo–reportage were also growing in popularity. The first of these, the *Illustrated American*, preceded the popular German publications by almost 30 years. Its first issue appeared on February 22, 1890. The stated aims were to develop the possibilities of the almost unexplored uses and work of the camera. That first issue met its goal: 75 photographs were used in the inaugural issue.

Thirty years after the *Illustrated American* broke ground, Henry Luce founded the magazine *Fortune*. Luce also believed in the concepts of attractively packaged photography and text. When he hired Margaret Bourke–White to work on *Fortune*, he told her of his plan: "In this new publication, words and pictures should be partners."[7] Luce continued in this vein when, in 1936, he instituted another magazine, one which is considered the masterpiece of the picture–story genre, *Life*.

Luce well understood the public's desire to *see* the news, then read a short caption, versus reading a lengthy story accompanied by a small illustration. This understanding was stated best by his prospectus for *Life*: "To see life, to see the world, to eyewitness great events; to watch the faces of the poor and the gestures of the proud; to see strange things—machines, armies, multitudes, shadows in the jungle and on the moon; to see man's work—his paintings, towers, and discoveries; to see things a thousand miles away, things hidden behind walls and within rooms, things dangerous to come to; the women that men love and many children; to see and take pleasure in seeing; to see and be amazed; to see and be instructed."[8]

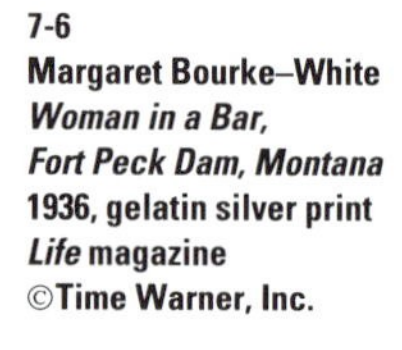
7-6
Margaret Bourke–White
Woman in a Bar,
Fort Peck Dam, Montana
1936, gelatin silver print
***Life* magazine**
©Time Warner, Inc.

John Billings was hired as the first managing editor. Although he had the assistance of researchers, designers, and subeditors, John Billings' task was enormous. As he described his job, "I will decide what I want, what to throw away, I will lay it out, I will place it in the magazine, I will decide how much space it should have. I will edit the copy that is written about the pictures. And on Saturday night at 6 o'clock, I will put on my hat and go home." The length of his workday was not dissimilar to that of the editors of the German magazines published during that same era.

Another tradition used in many of the German illustrated magazines was carried over to the pages of *Life* magazine—that of the picture story. Margaret Bourke–White, a well–known industrial photographer originally hired to work at *Fortune*, was transferred to *Life* in order to photograph for its first issue—datelined November 23, 1936.

Bourke–White was assigned to travel to a small town in Montana to photograph the building of a Works Progress Administration (WPA) dam. She was chosen for the task as the dam was an industrial structure and well–suited to Bourke–White's particular style of journalism. When she returned to the offices of *Life* with her assignment complete, she caught her editors off guard. As they wrote in their introduction to the first issue, "What the editors expected—for use in some later issue—were construction pictures as only Bourke–White can take them. What the editors got was a human document of frontier life which, to them at least, was a revelation."[9] That first issue published in November 1936 featured Bourke–White's images designed into *Life*'s first picture essay. Bourke–White's photograph of the dam was on the cover. In addition, nine pages of the magazine were devoted to photographs of the dam workers, their families, and their way of life in a small town on the Montana plains. Captions were minimal. The pictures were the story.

During the years that followed, other magazines followed *Life*'s approach to photojournalism and the picture story. *Look* appeared on the stands in 1937 as a magazine dedicated to feature stories. In Europe, the *Picture Post*, *Paris Match*, and *Der Spiegel* were published to answer the demands of millions of readers who believed that—in the words of John Szarkowski—"if one picture is worth a thousand words, ten pictures would make a short story."[10]

Alfred Eisenstadt was another photographer plucked from his position as an AP photographer in Germany to work for the pages of *Life*. In the 40 years of their association, Eisenstadt had over 1,000 feature stories published in *Life*. Some were gestural portrait studies, exemplified by his series of English notables published in January of 1952. Others hearkened back to his previous experiences in Germany. They centered around conflict. Most notably he photographed the effects of the war in Ethiopia.

W. Eugene Smith was another photojournalist whose work was featured in *Life*. Initially a part-time employee, Smith was hired through his work at the BLACK STAR picture agency. His assignment was to cover national "human interest" events. In his words, "I joined *Life* the day they ran out of other photographers because Bourke-White, Carl Mydans, and their other aces were running around the world with the U.S. Fleet."[11] It was evident that Smith was not altogether happy in those first years at *Life*, as he resigned within two years in order to cover the Navy's aircraft carrier and fighter-bomber consortia for Ziff-Davis (a rival publishing house). As Smith recounts; "I resigned from *Life* when...I became disturbed and fed up with assignments such as "Sadie Hawkins' Day" and the "Butler's Ball," things like that...I had made a mistake by resigning....*Life* was really the only outfit to work for if you covered a war. They had the greatest freedom, the greatest power, and the best expense accounts!"[12]

After covering the Pacific fleet during the war and sustaining serious injuries at Okinawa, Smith returned to the United States. After some hesitation, he again joined the ranks of *Life* staff photographers. Between 1946 and 1952, Smith left an indelible imprint on the profession of photojournalism with his picture essays. "Country Doctor," "Nurse Midwife," and "Spanish Village" were among the strongest of his more than 50 assignments completed for *Life* during this period. The next two years were fallow for Smith and in 1954 he terminated his relationship with *Life* after a bitter dispute concerning publication methods.

7-7
Alfred Eisenstadt
Dr. Joseph Goebbels
1933, gelatin silver print
***Life* magazine**
©Time Warner, Inc.

Those first ten years of *Life* (1936–1946) coincided with many global conflicts. Mindful of its stated aim to "see strange things—machines, armies, multitudes,"[13] *Life* sent a coterie of photographers to the front lines to capture the drama and tragedy of war. In World War II alone, *Life* had 21 full-time photographers covering the scope of international conflict (Chapter 11). Alfred Eisenstadt concentrated on the individuals who made news by making war. Margaret Bourke-White immortalized those victims whose mortality served as an icon for the tragedy of war (Figure 11-10).

Later, during the Korean War, and subsequently the Vietnam conflict, *Life* continued to provide the English-speaking populace with unforgettable battle coverage. During this time, such photographers as David Douglas Duncan and Larry Burrows continued the visual methodology of Smith and Bourke-White. Another young photographer, Bill Eppridge, worked for *Life* first as a part-time "stringer" and then, in 1965, as a full-time staff member. Almost immediately after his hire, he was assigned to cover the United States Marines in Vietnam. Subsequently, he was assigned by *Life* to cover a rising young American political candidate, Robert Kennedy.

Life and Politics

The Kennedy family became a symbol of America during the 1960s. *Life* magazine assigned particular photographers to cover as many aspects of life in "Camelot"* as possible. It is partially through these images of the Kennedys, captured by Eppridge and others, that we more fully realize the impact of photojournalism on our respective psyches. We recall memorable images of John F. Kennedy and his rise to political prominence in the highest elected office; photographs of his subsequent assassination and funeral; the political rise of Robert Kennedy (as photographed by Eppridge and others); his tragic assassination as covered by Eppridge in the kitchen of a Los Angeles hotel. Eppridge was with Kennedy at the moment of the shooting. Although in shock, Eppridge remembers saying to himself, "O.K. you've had ten years training for a moment like this...now go to work."[14]

His reaction was appropriate for a professional photojournalist, whose charge is to be the "eyes" for a public that desires information. This can mean countless hours on convention floors, photographing the seemingly endless hours of harangue in order to capture one image that might illuminate the machinations of the democratic system of government. It can take the unfortunate form of catching violence aimed at public figures. Or, it can be a "photo opportunity"—a term coined by President Nixon's press secretary, Ron Ziegler, in 1969 to describe all–too–brief camera sessions with the President or his advisors, normally on the politician's terms.

The Decline and Resurrection of *Life*

The magazine *Life*—with its sister publications *Look*, *Parade*, and the *Saturday Evening Post*—chronicled the full gamut of the American experience from the mid–1930s to the early 1970s. It seemed, at the time, that the publishers of these magazines had found a perfect formula for a periodical with eternal life. However, in December 1972, an announcement was made that *Life* was going to cease publication. Likewise, the death knell had sounded for many other magazines. *Look*, *Parade*, the *Saturday Evening Post*, and *Collier's*, published as long as possible, then were forced to write their own obituaries. The advent of video media had taken its toll.

By the late 1950s, television had gained a stranglehold on the imagination of the American public. This video force had an adverse, twofold effect on picture magazines. First, as commercial industry realized that the public consumer was watching ever–increasing amounts of television, they switched their advertising dollar to the new medium. Rather than produce advertisements that would "sit" on a printed page, industry realized that the product could be presented in a more attractive and livelier manner in the form of "commercials." Second, magazines required a semblance of literacy. Television, on the other hand, promised pure "entertainment." People were increasingly interested in this passive form of gathering information. They were also becoming more accustomed to instantaneity, whether it be toward late–breaking news or purely popular information.

However, in the late 1970s, *Life* and the *Saturday Evening Post* resurfaced with new outlooks. As monthly magazines, they now rely on popular reportage, sleeker design, and very high quality printing. In the case of *Life*, entire issues are printed in the expensive four–color system. The cover price reflects this extravagance. However, as a monthly (as opposed to a weekly) publication, it has found a continued niche in the field of magazine reportage.

Color Reproduction in Photographic Journals

Life was certainly not the first publication to realize the advantage of color illustrations. As early as 1840, the Currier and Ives printing company used the chromolithography process of four–color printing. It was, however, slow and expensive, requiring a separate lithographic stone for each color reproduced. Because of the labor and expense involved, most of the lithographs produced by Currier and Ives, as well as the color plates in popular fashion magazines of the mid–1800s, were hand colored and not mechanically printed. The illustrations were stenciled in color, in an assembly-line operation. One person painted the red parts

**Camelot* as described in Arthurian legend, was a land where perfection in all things was normal. The term was used to describe the general mood of the early 1960s, as symbolized by the politics and life–styles of the Kennedy family.

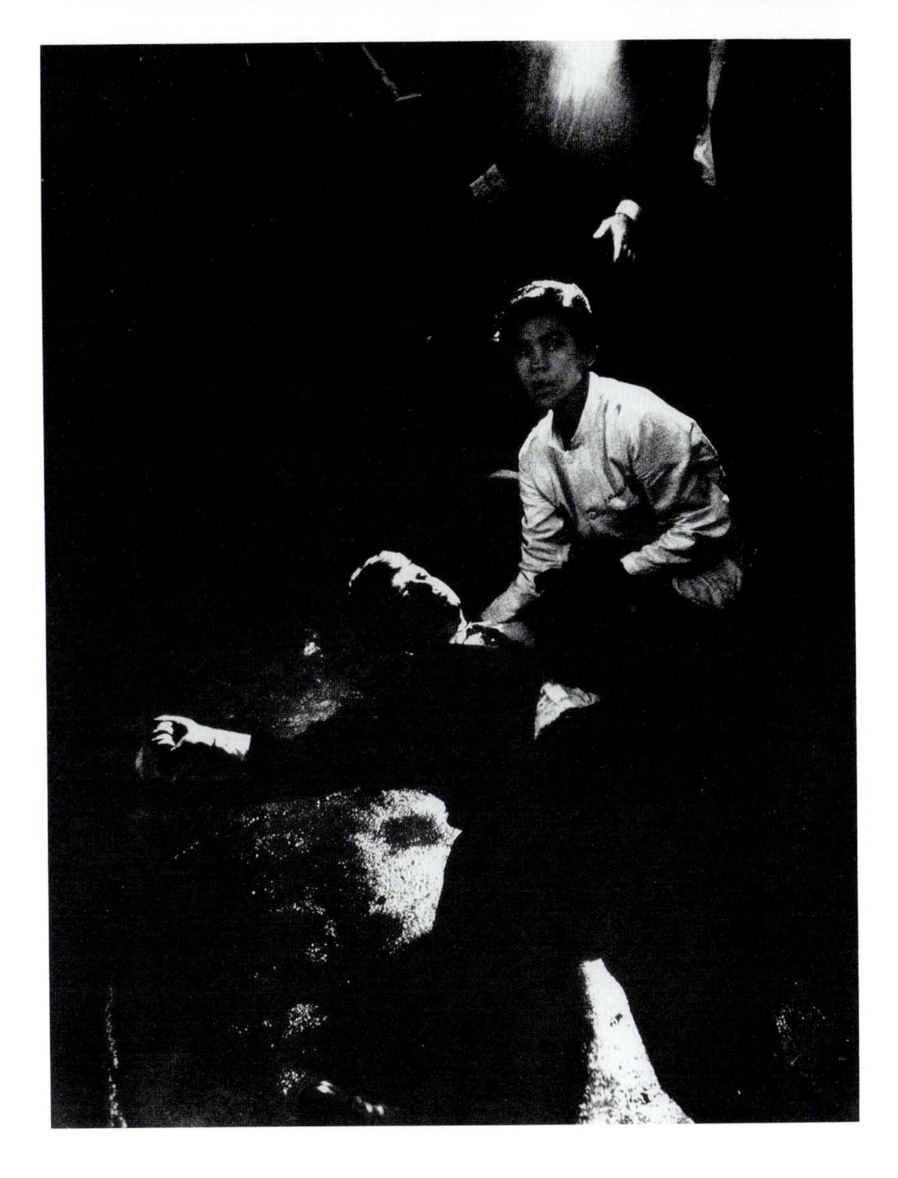

of the engraving, another, the blue parts. This method produced lively, but quite expensive, illustrations. The rotogravure method of printing (cousin to the photogravure), was another method that attempted to bring at least a tinge of color to the printed page. *National Geographic,* as early as 1907, employed the Autochrome system explained in Chapter 2. In fact, in many respects, *National Geographic* was the pioneer in color photographic reportage.

In 1889, *National Geographic*'s fourth issue featured pale color reproductions of views of Nicaragua. For the next 40 years, the magazine alternately used the Autochrome process, Findlay color, and the Agfacolor processes. By the early 1930s, a new process, Dufaycolor, had devised a mosaic screen of crisscross color on film instead of glass. However, the breakthrough in color photojournalism came with the introduction of Kodachrome in the 1930s. The emulsion was fast, permitting photography of human subjects. It was also designed for use with "hand cameras," essential for quick shooting. As the film's emulsion was on rolls, sequential images could be made in moments. Also, the roll film was naturally smaller, lighter, and less fragile than the traditional glass plate.

New Technologies—1960 to 1990

The 1930s were an age of evolution for photojournalism. Along with Kodachrome, the decade also ushered in widespread photojournalistic use of the hand–held 35–mm camera. In addition, Wirephotos and phototelegraphy were in wide usage between the mid–1920s and 1930s. By 1960, a new form of communication emerged—using the laser beam.

Lasers form a beam of light that can travel for long distances without "fanning out" or growing weak. Unlike the Wirephoto that was invented in 1924, the laser beam produces an extremely high quality transmission of "wire" pictures. It has been described as "a quantum leap...pure photographic quality for outstanding reproduction...of unmatched sharpness and clarity, all done automatically."[15] In 1974, the Associated Press committed itself to this new technology by replacing all of its Wirephoto machines with Laserphoto receivers and transmitters.

Another technological advance of the early 1970s involved the Unifax machine of the United Press International. It used an electrostatic recording process to transceive (transmit and receive) photographically generated imagery. Used for photojournalistic applications, as well as weather map reception, it is also invaluable in law enforcement assistance. The sophistication of the Unifax machine has grown since its introduction in 1972. By the end of the 1970s, computers had enhanced the capabilities of this and most of the other machines involved in photojournalistic publication.

The "still–video" camera of the 1980s presents a further evolution in photojournalism through the use of computer technology. With this camera, it is possible for editors to have—on their desks and ready for press—images shot minutes earlier in another part of the country. The technology is so revolutionary that some question exists as to whether it is truly a photographic process.

7-8
Bill Eppridge
Robert Kennedy Assassination
1968, gelatin silver print
***Life* magazine**
©Time Warner, Inc.

There is no need for film processing. In fact, there is not even any use for film. The camera records light on a silicon chip. This chip contains 600,000 photoreceptors that record the scene on elements called pixels. The image is converted by the chip to electrical impulses and is transferred to a small computer disk. Each disk holds between 25 and 30 "shots." When a disk is full of visual information, another is substituted in the camera. When the shoot is finished, the photographer plugs his disk into a portable monitor–transmitter and can edit the images, if required. If the edited images need to be sent to another location, the transmitter is simply plugged into a phone. Electromagnetic data is readily translated by the telephone to a transceiver at the publication office. The pictures are transmitted at the speed of sound.

Computers are also routinely used in electronic darkrooms. Computers receive transmitted imagery and are able to store, retrieve, or transmit it at a later date. An editor using a computerized, electronic darkroom has myriad choices once the electronic image is pulled out of its memory and flashed onto the screen. He can crop it, enlarge it, darken it, change its tonal relationships—even retouch it.

Controversy has recently erupted concerning the computerized retouching of electronic images. As the computerized darkroom has the capability to meld various portions from different images, it is possible to create a fictional photograph that has every semblance of reality. Normally, this type of electronic "plastic surgery" is used only to improve composition. However, it is possible to construct a fictional photograph. For example, world leaders could be portrayed as attending a summit meeting when no such event transpired.

Today's Photojournalists

Today's photojournalists cover a wide spectrum of events. They can have a national "hard" news orientation, presenting important global events, or they can choose "soft" news, such as feature stories on current trends. They may be "paparazzi" (from French, meaning "scribbler"), or they may choose to shoot more serious imagery.

A paparazzo is nearly always involved in shadowing current society figures, snapping photographs, and selling the images for use in popular weekly magazines, such as *People*, although newspapers are also willing buyers. It is a somewhat dangerous occupation, as most of the subjects are not willing to be followed 24 hours a day. Actor Sean Penn has been convicted of manhandling various paparazzi. Jacqueline Kennedy Onassis (widow of John F. Kennedy) has been plagued by paparazzo Ron Galella so often that she requested—and received—a federal court order that limits Galella to shooting at a distance of over 50 feet from his favorite subject.

The paparazzi are at one end of the spectrum of those who photograph famous people for publication. At the other end are more aesthetically–oriented photographers who contract to photograph portraits of famous people for publication. As such, they must be considered photojournalists, although many of these same photographers are discussed in Chapter 6.

In the 1930s, magazines started hiring photographers who specialized in portraiture. These magazines wished to engage in a civilized method of capturing the likenesses of famous subjects. Edward Steichen photographed the elusive Greta Garbo for a Condé Nast publication. Irving Penn and Richard Avedon worked for many current feature magazines, most notably *Life*. Annie Liebowitz was originally hired to work for a music industry publication, *Rolling Stone*. Her assignment was to capture the essence of figures at the forefront of popular culture. Her style, like many of the aforementioned photographers, was unmistakable.

Some photojournalists are stringers, (who shoot, then attempt to sell their imagery), others are contracted by particular publications to cover specific assignments. Occasionally, the more successful photojournalists unite to create an entity that allows them the freedom to shoot any current subject—regardless of whether or not there is a precontracted agreement with a publisher. MAGNUM was one of the first of these collaboratives. It has a finite group of elected members, all of whom donate funds to a "pool." This money is then used for travel or for any number of professionally related necessities. MAGNUM was founded by Robert Capa (Chapter 11), Henri Cartier-Bresson (Chapter 10), Chim (David Seymour), and George Rodger in Paris in 1947 and opened a New York office within a year. Around the time MAGNUM was founded, other photographic collaboratives emerged. One of the largest of these collaboratives is BLACK STAR, started in Germany in 1936.

Although not currently a member of a photojournalism collaborative, Mary Ellen Mark is an example of a new breed of photojournalist. She has earned a master's degree in the specific field of photojournalism. Like Jacob Riis and Lewis W. Hine, she chooses to witness and expose crucial humanistic events of her time. She was one of the first photojournalists to cover the Civil Rights struggles of the 1960s. Later, again like Riis and Hine, she centered her work around people on the margins of society. Portfolios, even books, have been devoted to her imagery of runaways, prostitutes, the homeless, and the mentally ill. Most recently, she has been working in Europe and the Far East, covering the effects of famine.

Eugene Richards also follows in the footsteps of Riis and Hine. Like Riis, he has worked as a newspaper photographer. Like Hine, he has worked closely with his subjects—once spending four years as a social worker in a poverty-stricken area of Arkansas. When *Life* magazine published its issue celebrating 150 years of photography (in 1988), Richards' image of a small black child holding a white doll was featured on the same page as Riis' photograph of a group of small children huddling in a doorway. Richards is also quoted in that issue as saying that he is compelled "to bring things to light. I am more concerned than ever about the way the world is going. I want to be involved in it." In the past ten years, Richards has documented such varied subjects as cancer, poverty, and urban hospital emergency rooms. His work is often collated into books, the most recent of these being *The Knife and Gun Club*. He is a member of MAGNUM.

Some photographers, as mentioned in Chapter 11, concentrate on global wars. Others function as local news photojournalists. Still others are employed by the United States federal government for the express purpose of capturing the President on film. The realm of photojournalism is vast. In the last 100 years, technology has made the job easier, but social issues continue unabated. Where there are people, there will be events. Where there are events of interest, there will be photojournalists.

7-9
Mary Ellen Mark
Streetwise
n.d., gelatin silver print
courtesy M. E. Mark/Library

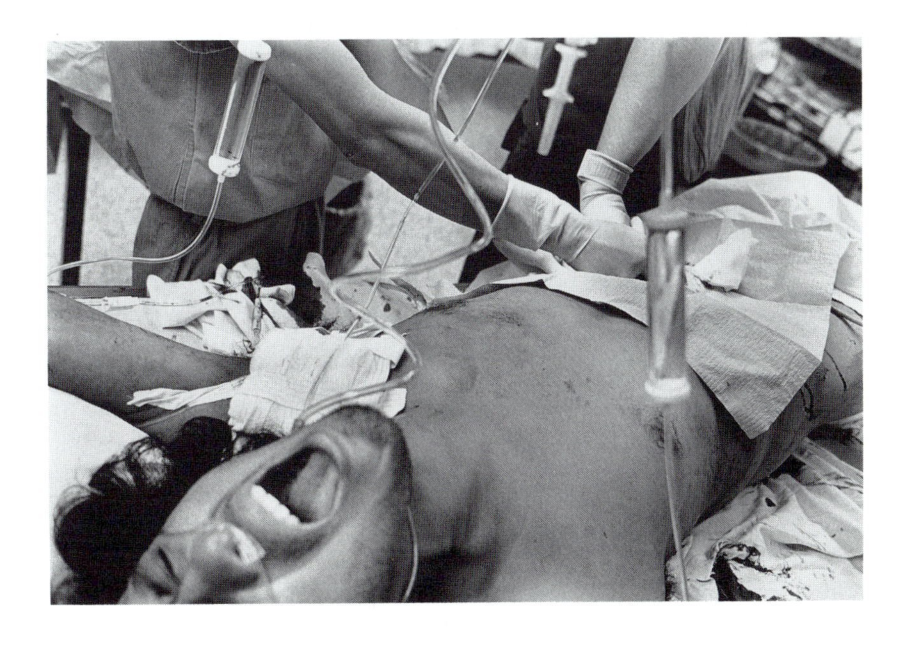

7-10
Eugene Richards
from *"The Knife and Gun Club"*
n.d., gelatin silver print
©Eugene Richards
courtesy MAGNUM

To ignore photojouralism is to ignore history.[16]
—Howard Chapnick
President, BLACK STAR

8-1
Clarence White
Ring Toss
1899, platinum print
Museum of Art, RISD

8

Alfred Stieglitz and the Photo–Secession

Look again! Is it not Alfred the Great? Not that Alfred who let the cakes burn. No; this is the Alfred who pulled the cakes out of the fire. The Photographic Cakes. Yes, dear; his hair *is* a little rumply, but if you were ringmaster in this show your hair would stand up too.[1]
—J.B. Kerfoot

In 1864, 25 years after the invention of photography, Alfred Stieglitz was born in Hoboken, New Jersey. His contribution to photography and modern art was symbolized by his birthdate—New Year's Day. His was a new vision—one that sifted, evaluated, and often discarded artistic tenets that (he felt) had outlasted their usefulness. He had his critics and the battles he fought were not fought without bloodshed. However, through strength of character, integrity, and an inordinate sense of self–worth, Alfred Stieglitz accomplished that which he sought—recognition of photography as a legitimate medium of artistic expression.

Snapshooters, Scientists, and "Artists"—1864 to 1890

In 1864, the year of Stieglitz's birth, the bulky wet–plate photographic process was still in use. However, within the next 20 years, technical advances and new manufacturing techniques permitted the production of thousands of inexpensive cameras that generated millions of photographic images. Naturally, most of these images were personal records of subjects that bore little interest to anyone but the photographer. Many of these negatives were underexposed, fuzzy, or suffered from any number of other technical maladies. Certainly, a vast majority of the pictures taken had no sense of aesthetic form or balance. They were simply snapshots.

Other photographers worked much more technically. These scientific photographers devoted their energies to photography as a device for recording data. Astronomers, medical scientists, and botanists used photography in their own particular fields of study. For them, photography was a means of problem solving and discovery. Although some of these scientific photographs had a particular fascinating beauty, they were still simply the result of technical exploration. They were not generated with any artistic intent.

Another subgroup of photographers were neither snapshooters nor scientists. These image makers wished to produce pictures of aesthetic substance. However, by and large, they were weekend artists. Joseph Pennell, a satirist and illustrator of the late 1800s, lampooned these amateur artist–photographers as those who labor at their desks all week, and then, when the half–holiday comes, seize their little black boxes, skip nimbly to the bus, and hurry to 'plant' their machines in convenient corners, and with the pressing of a button, create a nocturne that shall rank with the life–work of the Master.

This tongue–in–cheek description of most "art" photographers was, unfortunately, quite fitting. However, their motives were pure and directed toward using this new medium to communicate aesthetic ideals. Unfortunately, the "established" art world scoffed at their efforts (Chapter 3). At that time, most artists considered photography to be a mechanical curiosity, a craft, or—at best—an aid to painterly study. This purview was painfully evident to "art photographers" when they pounded vainly on the doors of galleries, begging to have their works shown. As late as 1890, photography had neither been shown on the same level as painting, nor had any photographs been directly purchased by a museum collection.

The Pictorial Aesthetic

Art photographers reacted to this situation by forming camera clubs or photographic societies. They reasoned that if their work could not be shown as equal to contemporary painting, they could mount their own, separate exhibitions. Normally, the clubs had relatively small active memberships, so a number of these groups organized joint exhibitions. These shows covered all motifs explored through the photographic medium and works were often displayed according to the unsophisticated system of subject matter. All marine views would hang in one area, another area would display figure compositions, and scientific views would hang separately from portraits.

Camera club members had a wide variety of interests, backgrounds, and experiences. This diversity prompted the societies to host lectures on varied topics. Featured speakers gave demonstrations pertaining to the newest equipment or the latest photographic process. If composition and design were discussed, the lecture would be based on the immutable aesthetics of the painterly establishment. Painting was considered "high art." Reason dictated that if painting represented the best of the art world, then the compositional rules and subject matter of painting must be translated photographically in order for this new medium to gain acceptance.

At that time, the painterly aesthetic tended toward soft, remote, pastoral subject matter. An object in a painting, according to nineteenth century art academics, should be subjectively presented. If a canvas were to be compared with a window it should have cloudy glass and reveal, not the view outside, but that of the inner landscape of the artist's soul.

"Art" photographers, attempting to mimic this aesthetic, were aided by writer–photographer Henry Peach Robinson (Chapter 12). In 1869, he published a veritable guidebook on aesthetic photographic principles. It was titled *The Pictorial Effect in Photography*. He used the word pictorial, because he believed that the best "art" photographs should resemble paintings, or pictures, in their subject matter and visual execution. Robinson illustrated a portion of the book with his own pictorial work, which closely followed the painterly aesthetic (Figure 12-6).

Stieglitz's Early Years—1881 to 1890

This was the state of the academic art world when Alfred Stieglitz was born to his German immigrant parents. Alfred was one of six children whose childhood was uneventful save for the Sunday salons that his father held for learned friends. Any number of political, artistic, and cultural subjects were discussed at these informal gatherings. The only taboo was the word "business." As Alfred's family was modestly wealthy, he was offered on his 17th birthday the chance to study in Germany. In 1881, he left for Europe to study mechanical engineering at the Berlin Polytechnic Institute.

Not long after his arrival, Stieglitz was strolling down a street lined with shops when he spotted a camera in a storefront window. He bought it on a whim and soon was spending more time planning images than machines. He transferred out of mechanical engineering and worked with Hermann Wilhelm Vogel, a world–famous photochemist and champion of pictorial photography.

8-2
Alfred Stieglitz
Sun Rays, Paula, Berlin
1889, gelatin silver print
IMP/GEH

Stieglitz's early photographic interests were oriented primarily toward purely technical concerns. His mission, he felt, was to render the "perfect" negative, which would evidence all proper tones of blacks and whites. He became obsessed with this mission until, one day, Vogel advised him that it was impossible to achieve absolutes and that many compromises were necessary when photographing. Although Stieglitz respected Vogel, he inflexibly refused to believe that integrity, whether it be photographic or personal, must involve compromise.

A number of Stieglitz's technical and aesthetic concerns are evident in Figure 8-2. This photograph is an intimate scene, a mood favored by the pictorial ethic, and shows a mastery of technical problems not often confronted by photographers of the late 1880s. Such dramatic contrast between highlight and shadow is difficult to record on film. However, Stieglitz, with his commitment to technical excellence, managed to overcome the problematic lighting situation.

Most of Stieglitz's early photographic work ascribed to the pictorial ethic and to the ideas espoused by Emerson of photography as a singular art form. When Stieglitz submitted the photograph, *A Good Joke,* to a contest held in Britain (for which Emerson was sole judge), he won his first photographic prize (two guineas and a silver medal). Although it was a relatively uninspired genre scene, Emerson remarked that of the thousands of photographs he had seen as judge, Stieglitz's entry was the only one that was truly spontaneous. After the contest, Emerson and Stieglitz began to correspond. They had both grown up in America and both were living in Europe, had overwhelming energy, and an abiding interest in photography as an art form.

Emerson's award to Stieglitz was the first of 150 medals and prizes Stieglitz was to garner. His work was exhibited regularly in Europe, during the 1880s, to a largely enthusiastic audience. It was, therefore, very difficult for Alfred Stieglitz to uproot himself and return to America in 1890. One of his sisters had died, and after repeated urgings from his mother, Stieglitz returned to New York.

Stieglitz and the Camera Club—1890 to 1902

New York in 1890 was much different from the city he had left nine years earlier. The economy was depressed and the ideals of profit and exploitation had replaced the ethic of fair play in American business. Stieglitz had a difficult time finding employment. His father suggested selling some of his photographs. Alfred was appalled at the suggestion. The elder Stieglitz finally situated his son in a photoengraving–printing business that had been started by two friends. While Alfred worked diligently for 5 years, he never became truly involved in the work. He resigned in 1895.

When Alfred Stieglitz returned to America, he was disappointed to note that while amateur photography clubs were flourishing, they had very little vitality. Their stagnation was in sharp contrast with the excitement existent in the European photographic scene. Having fulfilled his original obligation to his family, he considered returning to Europe until a friend challenged him to try one of the new hand cameras that had recently been introduced to the market. These cameras used 4 x 5–inch plates. After borrowing one from a friend, Stieglitz began to roam the streets of New York. This marked the beginning of a 40–year photographic treatise involving Stieglitz and America's busiest metropolis.

Stieglitz was never noted for taking advice or for blindly following accepted patterns of behavior. When he started using the 4 x 5–inch camera, he realized that his methodology was at odds with most contemporary photographers. First, the 4 x 5–inch hand camera was scorned by "serious photographers" as a machine unworthy of producing art. Second, his choice of urban subject matter was at variance with the tenets of the established pictorialists. Third, when he decided to turn his underexposed negatives from these forays into lantern slides (or positive slide–show–type images), his colleagues were astounded. Lantern slides had historically been used either for entertainment purposes or as a method of illustrating scientific lectures. Emerson had dismissed lantern slides as "toys," further stating that they had no place in art. Stieglitz strongly disagreed. To demonstrate the implicit beauty of lantern slides, he assembled two slide demonstrations of the process and its value to his art. He exhibited these to the Society of Amateur Photographers in 1896.

This society, or "camera club," was one of the most active in New York during the 1890s. Although Stieglitz considered their sense of aesthetics unsophisticated, they did have one asset he could not dismiss—a photographic journal titled *American Amateur Photographer.* Upon joining the group, he immediately volunteered to edit this publication. *American Amateur Photographer* was the perfect vehicle to push for a stronger American aesthetic in photography. In his usual forceful manner, he completely redesigned the magazine and refused to publish works by a number of established photographers. While he personally ascribed to Emerson's aesthetic of naturalism, as an editor he favored no particular style. He published the freshest and most dynamic work submitted and harshly criticized mediocrity and clichéd subject matter. His rejection slips scandalized his publisher. His comments were terse and unflinching. Most were simple and pointed: "Technically perfect; pictorially rotten!"[2]

During this period, Stieglitz continued to exhibit his work in Europe. The climate for photography was much further advanced in Europe than in the United States. However, rather than simply depend on recognition from his foreign colleagues, Stieglitz hoped to convince fellow Americans of the importance of a national photographic identity. In 1895, he pleaded that "We Americans cannot afford to stand still; we have the best of material among us, hidden in many cases; let us bring it out...let's start afresh with an Annual Photographic Salon to be run upon the strictest lines."[3]

His ideal for an annual salon was not to be realized during his tenure with *American Amateur Photographer*. His relationship with the publisher had been shaky for some time and when in 1896, the publisher printed a photograph without credit to the artist, Stieglitz resigned his editorship. He still retained his membership in the Society of Amateur Photographers, however Stieglitz became aware of a mounting problem at the Society. It centered around debts that the Society continuously incurred. The membership simply could not afford to support all the programs demanded by their enthusiasm toward the medium. However, a sister club, the New York Camera Club, had excess funds, but a lackluster membership. Within a few months, the Society of Amateur Photographers merged with the New York Camera Club to become, simply, The Camera Club. Stieglitz was elected vice president and chairman of the publication committee. His condition for accepting the editorship of their existing leaflet publication, *Journal,* was that he be given "unhampered and absolute control of all matters, direct and remote, relating to the conduct"[4] of this quarterly.

While the initial issue of the new publication, renamed *Camera Notes*, resembled *American Amateur Photographer*, the quality of the illustrations immediately distinguished it from its predecessor. Each issue contained two hand–pulled photogravures (an exceptionally fine printing process) personally chosen by Stieglitz. Not content simply to have dynamic visual examples of pictorialism in *Camera Notes*, Stieglitz also expanded its critical base. He contacted artists, historians, and critics to write for this publication. By 1902, Stieglitz was in contact with, and publishing, a coterie of literate artists who represented excellence in their chosen fields.

In addition to his exhaustive work with *Camera Notes*, Stieglitz understood the importance of American pictorial photography exhibitions. However, he had very definite ideas concerning the structure of these "salons." First, they should have little in common with the joint exhibitions that had been organized between New York, Philadelphia, and Boston between 1884 and 1894. Second, membership in a club or society would not gain immediate acceptance for work to be hung and judging should be very strict. Third, no medals or prizes should be awarded. Stieglitz felt that the acceptance and hanging of a picture should be honor enough for the photographer.

In 1896, the Camera Club of the Capitol Bicycle Club of Washington announced a two–part exhibition. Class A (the Salon) was earmarked to show "high art." Class B was limited to amateurs. The selection of imagery to be hung was the responsibility of the artists. When it opened, it so impressed the Director of the United States National Museum that he purchased 50 prints for $300. It was the first recorded museum purchase of "art" photographs. The status of photography had been elevated by this purchase, but Stieglitz was not pleased with the exhibition. He did not participate, disagreed with the awarding of prizes, and was appalled at the lack of aesthetic standards. Of the 565 photographs submitted to the exhbition hanging committee, 485 were hung for display.

Two years later, in 1898, the Philadelphia Photographic Society announced the organization of another photographic exhibition. Their stated claim was to hang "such pictures produced by photography as may give distinct evidence of individual artistic feeling and execution."[5] It was a tremendous effort. Most of the images hung at this first Philadelphia Salon were by Americans, although invitations to exhibit were sent internationally. The 1898 Salon was so well received that it became an annual event. The first three Philadelphia Salons were wildly successful, although the strict standards of judging alienated many estab-

8-3
Alfred Stieglitz
The Terminal
from *Camera Work*, 1911
1892, photogravure
MOMA

lished photographers whose work had been rejected. Many had won awards for particular "masterpieces" and had seen those same images rejected by the Philadelphia hanging committee.

By the fourth annual Salon, these malcontents had joined together and pressured the judges to institute "broader" lines of acceptance. The result was disappointing. *Camera Notes* commented on the Exhibition of 1901. It was critical of the new, lower standards and closed its review thus: "The pictorial photographers of the country will (must) now form their own organization, and hold their own exhibitions where the best interests of pictorial photography will be more faithfully guarded and consistently served."[6]

The battle lines had been drawn. The mediocre photographers had strength in numbers. Stieglitz and a few colleagues had strength in their own integrity of vision. While editing *Camera Notes,* Stieglitz had come under increasing pressure to bow to the mediocre and include work he considered unacceptable. As vice president of The Camera Club, he had similar pressures from members. They wanted him to hang banal photographic work on the walls of the club. Stieglitz was unmoving in his rejection of the mediocre. In fact, Stieglitz hung work of photographers who were not even affiliated with the club, while snubbing most members' entreaties for gallery exposure. When not working on *Camera Notes* or other club business, Stieglitz was busy packaging exhibitions of nonmembers' work. He sent these groups of images to colleagues around the globe, attempting to champion the best in pictorial photography.

His actions alienated a majority of The Camera Club membership. The final split between Stieglitz and The Camera Club occurred in 1902. A new administration had been elected, which found fault with Stieglitz's methods of editing *Camera Notes*. Stieglitz resigned the editorship out of self-respect and a belief that, after five years of hard work, his policies should be inviolate. His final issue was Volume VI, Number 1. Although a new editor was hired, the vitality of the periodical was gone. The publication of *Camera Notes* ceased three issues later. Although Stieglitz kept his membership in The Camera

Club until he was asked to resign in 1908, he was never again an active member. In fact, Stieglitz deliberately separated himself from that organization for which he had worked so diligently in order to more fully promote selected photographers he had championed.

In 1902, Charles McKay, art critic for the *New York Times* and a friend of Stieglitz's, suggested that he hang an exhibition representative of these photographers. McKay's powerful position in the art world allowed him to request space for the exhibition from the National Arts Club. They offered Stieglitz a room at their location for the exhibition and Stieglitz accepted. Controversy arose, however, when it came time to title the exhibition. McKay suggested naming it "An Exhibition of American Photography Arranged by The Camera Club of New York." Stieglitz refused, remarking that The Camera Club was against everything he believed about photography.

The Photo-Secession—1902 to 1917

Stieglitz preferred to title the show "An Exhibition of American Pictorial Photography Arranged by the 'Photo-Secession'." When the chairman of the Exhibition Committee for the National Arts Club asked Stieglitz, "What is the Photo-Secession?," Stieglitz answered,"Yours truly for the present...In Europe...the moderns call themselves Secessionists, so Photo-Secession really hitches up with the art world."[7] He was referring to the German and Austrian *Sezession* painters who had reacted against the established rules of salon art. Thus, Stieglitz, in naming the exhibition, had made his motives clear. He had decided to align photography with a broader concept of art than that of simple pictorialism. He was comparing himself to the *Sezession* group, as well as to the Linked Ring Brotherhood of photographers who had seceded from the established Royal Photographic Society due to artistic differences. He intended to align himself and selected colleagues with other disaffected groups of artists working in diverging art forms.

Stieglitz arranged the exhibition and organized the artists into a loosely formed group—The Photo-Secessionists—in mid-February 1902. Twelve founders were chosen and membership was divided into fellows and associates. The stated aims of the group were to advance the very best in photography and its practioners, and to hold sporadic exhibitions not limited to American work and, further, not even limited to photography.

The first exhibition was held at the National Arts Club only weeks after the group officially formed. The show was open to the public for 17 days and was well received by some critics, who recognized a new and sophisticated photographic aesthetic. Others condemned the works of these featured pictorialists as imitations of bad paintings. However, when a selection of these works was sent to Europe, it was universally acclaimed. Ernst Schul, a German critic, hailed the photo-secessionists as "the most modern of all, yet the most sure and reposeful...they reach a consciously selected goal with the calm of perfect deliberation, like the hunter who, with cool and deadly aim, reaches his prey."[8]

Concurrently, Stieglitz decided to publish an illustrated quarterly magazine devoted to photography. He vowed that it would "owe allegiance to no organization or clique" but paradoxically would be "the mouthpiece of the Photo-Secession."[9] Of course, Stieglitz would edit and control the publication. When he initially announced this forthcoming quarterly, he must have been gratified by the response. Although only sketchy introductory information was given about *Camera Work,* Stieglitz soon had 647 subscription requests.

In the first years of *Camera Work*'s publication, Stieglitz requested work from many former colleagues who had been featured in *Camera Notes*. *Camera Work's* format was somewhat consistent with that of *Camera Notes*, though more elaborate. Each issue was printed in an edition of 1,000 copies. Each (except for one) contained a hand-tipped (or nonmechanically placed) gravure portfolio by a single artist. Several additional photographic plates were featured in each issue. Edward Steichen, in an unprecedented gesture, personally hand-toned each of his single photogravures included in the 1,000 copies of Issue 19.

8-4
Gertrude Käsebier
Autumn
1903, photogravure
Museum of Art, RISD

Edward Steichen and Gertrude Käsebier

The devotion that would lead an artist to tint endless numbers of published photographs was indicative of the devotion Steichen felt for Stieglitz and his aesthetic ideals. Edward Steichen was, like Stieglitz, of a foreign–born family. He grew up in the Midwest and was educated in art while working as an apprentice at a lithographic company. Although trained as a painter, he began photographing at the age of 17 (approximately the same age as was Stieglitz when he began). Within a few years, he was exhibiting his works. His images caught the eye of both Clarence White and Alfred Stieglitz when they judged the Chicago Photographic Salon of 1900. Later that year, Steichen stopped to meet Stieglitz in New York while on his way to Europe. Stieglitz promptly bought three of Steichen's works—for $5 apiece.

Steichen stayed in Europe for a few years, continuing to paint as well as photograph. He was elected a member of the British Linked Ring, as well as a founding member of the Photo–Secession (in absentia). As a painter–photographer, he refused to recognize artificial delineations between art forms. He felt that art was either good, fresh, and modern or it shouldn't be acknowledged. It was the same credo that Stieglitz espoused. The ideals of Steichen and Stieglitz were so similar that the two became natural collaborators upon Steichen's return to New York in 1902.

Steichen soon became acquainted with the other local members of the photo–secession group and acted as Stieglitz's assistant in designing the cover and typography of *Camera Work*. The first issue was devoted to the work of Gertrude Käsebier.

Käsebier, like Steichen, initially had been trained as a painter. Her chosen subject was the portrait. However, soon after she began her formal studies (at the age of 36), she discovered photography and laid down her palette. Nine years later, she opened a professional photography studio—specializing in portraiture. Immediately her work was recognized for pictorial excellence. Around this time (1897), she met Stieglitz. They formed a strong bond and she was invited to become a founding member of the Photo–Secession, in addition to her membership in the Linked Ring. Although Käsebier ultimately broke with Stieglitz, hers was the strongest feminine voice in the photo–secession movement.

It is interesting to note that two of the twelve founding members of the Photo–Secession were women—a much higher female–to–male ratio than existed among photographers in general during that period. As mentioned, the premier issue of *Camera Work* was devoted to the work of Käsebier; it also featured an article, "Signatures," written by the other female founding member, Eva Watson–Schütze. An article was also included about Käsebier's professional life and was written by a woman, Frances Benjamin Johnston. From these observations, it may be safe to assume that sexism was nonexistent among the photo–

secessionists. In fact, Käsebier was dubbed "The Madonna of the Lens" when a writer for *Camera Work* satirically formed "King Arthur's [Stieglitz's, of course] Round Table."[10]

The second issue of *Camera Work* was devoted to a study of Steichen's photographs. Steichen's success as a painter bolstered his credibility as a photographer. His paintings and photographs shared many common elements—among them a mysterious, ethereal quality. His work in both media leant credence to the idea of photography as a fine–arts medium. Steichen, never favoring one medium of expression over another, had, at one point, simultaneous New York exhibitions—one of photographs and one of paintings.

The Little Galleries of the Photo–Secession and "291"

Steichen's life had become very active since his 1902 arrival in New York. He was producing his own photographic work, assisting Stieglitz on the design of *Camera Work*, and helping to package the photo–secession work that traveled to many parts of the world. The collections of work (the photo–secessionists always exhibited collectively) were widely requested by national and international galleries and museums. However, since their first exhibition at the National Arts Club, their work had not been featured in New York. While Stieglitz's founding notion for the photo–secession was "to hold...at varying places, exhibitions"[11] and initially was against having a permanent exhibition space, he eventually acceded to the idea that a location in New York should be found to showcase the work of the secessionists, among others.

In 1905, Steichen invited Stieglitz to his home at 291 Fifth Avenue. The two rooms next to his own had been vacated and Steichen suggested that the group rent them as an exhibition and meeting place specifically for use by the Photo–Secession. Although initially reluctant, Stieglitz agreed and announced to the members, in a letter dated October 14, 1905, that quarters in New York had been secured. The letter further stated that there would be continuous exhibitions to run for 2–week periods, of anywhere from 30 to 40 works of art (not necessarily photographs); that the work shown would be international; and that it would be open to the public.

8-5
Edward Steichen
Rodin—Le Penseur
1902, gravure
MOMA

Steichen was given the task of transforming the two rooms into a showplace for avant–garde art. He covered the walls with a bleached natural and olive–colored burlap, the ceiling was painted a deep creamy gray, and the woodwork and moldings were white. The first exhibition, 100 photographs by members of the Photo–Secession, opened to critical acclaim on November 24, 1905. Two weeks later a French exhibition opened featuring works by Robert Demachy (Figure 12-7). Later, works from Britain, Germany, and Austria were hung, along with work by the American secessionists. Critics hailed the Little Galleries as the place where the battle for recognition of pictorial photography as an art form had been won. The first few seasons were phenomenally successful.

By 1906, both Stieglitz and Steichen realized that it could be advantageous to the medium of photography if it were to be shown in direct comparison to other visual media. Therefore, in January of 1907, the Little Galleries mounted a showing of nonphotographic images (wash drawings) in conjunction with photographs. Concurrent with Stieglitz's decision to show different media was his decision to change the name of the gallery. The Little Galleries for the Photo–Secession was too limiting for Stieglitz's evolving aesthetic taste. Henceforth, it was defined simply by the number on the building, "291."

In answer to questions and rivalry among certain members of the Photo–Secession, Stieglitz attempted to explain his rationale of changing the gallery name, as well as his decision to show non–photographic work. He stated that "The Secession Idea is neither the servant nor the product of a medium. It is a spirit."[12] In order to nourish that spirit, Edward Steichen returned to Europe to seek new work by European artists. On his earlier travels, he had developed a friendship and correspondence with the French sculptor Auguste Rodin. When Steichen visited the artist in France, he selected and sent a number of Rodin's drawings back to Stieglitz. They were promptly exhibited. The following April, the drawings and lithographs of Henri Matisse were featured. Within months, more bold new work by such recognized artists as Henri Toulouse–Lautrec were exhibited at 291.

The New York critics responded to the "new" art with suspicion and hostility. They referred to Matisse's sketches as filled with "subterhuman hideousness."[13] Stieglitz answered their criticism with the statement that "The New York 'art world' was sorely in need of an irritant and Matisse certainly proved a timely one."[14] (Figure 3-4).

The Buffalo and Armory Exhibitions, Modern Art, and Dada

Never daunted by criticism, Stieglitz continued to mount exhibitions by ascribers to the cubist movement: Van Gogh, Picasso, and many others. By 1910, even *Camera Work* had begun to publish nonphotographic works. Although Stieglitz believed in presenting excellence in all media, he remained recognized as the champion of fine art photography. In 1910, he was invited to arrange an international exhibition of pictorial photography at the Albright Art Gallery in Buffalo, New York. Stieglitz demanded and received complete control. Stieglitz and his colleagues (not limited to photographers), covered the museum walls with cloth before they hung the 584 chosen photographs. Sixty–five artists were represented from four nationalities. Each artist was encouraged to show numerous pieces of work. By doing so, individual artistic development in the medium could be demonstrated. The show was a resounding success. The museum bought 15 prints and noted that they would set aside a permanent exhibition space for photographs.

Oddly, the success of the Buffalo exhibition signaled the beginning of a slow downward spiral for pictorialists. With the decisive recognition of photography as a fine art, much of the secessionists' original sense of purpose had been answered. Stieglitz himself was leaning toward a "straighter" or "purer" approach to the medium in his own work, as well as in some of the works he chose to show in Buffalo. The Linked Ring had dissolved. Furthermore, *Camera Work* was almost exclusively a forum for modern painting.

Three years later, in 1913, the Association of American Painters and Sculptors decided to hold an international exhibition in the Armory of the 69th Regiment in New York. Although

Stieglitz did not participate, he was actively consulted by exhibition coordinators and ultimately wrote an article, published in the *New York American,* which urged the public to attend and to respond. Concurrent with the armory show, for the first time in 14 years, Stieglitz hung a one-man exhibition of his own photographs. The work was startlingly spare, unromantic, and pure. That year he wrote that "Photographers must learn not to be ashamed to have their photographs look like photographs."[15] It was a definitive statement that signaled the end of his sympathies with the pictorialists.

Between 1913 and 1917, only one photographic exhibition was held at 291. In 1911, a double issue of *Camera Work* was devoted entirely to modern art. Stieglitz had purposely omitted any photographic work. Half of its remaining subscribers, bewildered by lack of even a single mention of photography, withdrew their subscriptions. Although in remaining issues Stieglitz did include photographs and critical writing about the medium, the bulk of each issue was devoted to articles by proponents of modern art and the new school of aesthetics, that originated in Zurich, Switzerland—the dada.

The dada ethic was a protest against the conventional, sentimental admiration of art. It represented a new form of secession from the established aesthetic, thus it was very attractive to Stieglitz. When the last regular issue of *Camera Work* was published in 1915, Stieglitz immediately began work on a new "little" magazine, *291*, edited by prominent dada artists. Its existence was announced in issue number 48 of *Camera Work* (published 21 months after number 47), along with a notation that a new gallery, The Modern Gallery, was open in New York. It mentioned that paintings of "the most advanced character of the Modern Art movement," as well as negro sculptures and Mexican art, would be featured. It went on to state that 291 would remain open, and that "The Modern Gallery is but an additional expression of 291."[16]

Alfred Stieglitz, who had once been sole champion of pictorial photography, was now the champion of modern art. While many were fearful that Stieglitz had turned away from photography, it was not true. He had simply recognized photography as a separate but equal entity. Doubters of Stieglitz's motives were, therefore, understandably relieved when, in 1917, Stieglitz featured the work of a relatively unknown, but powerful young photographer—Paul Strand.

8-6
Paul Strand
Abstraction
1915, from *Camera Work*, June 1917, photogravure
MOMA

Paul Strand

Paul Strand was born in 1890. He was educated at the Ethical Culture School, where he studied photography with Lewis W. Hine (Chapter 4). One day, Hine's class took a field trip to 291. It was Strand's first exposure to both Stieglitz and the art he championed. Although Strand was initially puzzled by the modern art displayed on the walls, he was intrigued and eventually became a frequent visitor to the gallery.

The exposure to modern art affected his photographic work profoundly. When he brought his portfolio to Stieglitz in 1915, Stieglitz described the work as, "brutally direct, pure and devoid of trickery."[17] It was the strongest praise Stieglitz was capable of bestowing. He immediately scheduled an exhibition of Strand's work at 291, and resurrected *Camera Work* for one final double issue in order to present Strand's work.

Strand's photographs were the antithesis of the pictorial tradition. They used a very "straight" photographic methodology. Their design and formal organization revealed an acute comprehension of modern art. Stieglitz was very pleased that he had finally seen a cohesive body of work that cogently melded photography and modern art.

Stieglitz's Later Years—1917 to 1946

After the June 1917 issue of *Camera Work* and the exhibition that had featured Strand's work, Stieglitz retreated to a more personal life. He closed the doors of 291 and ceased publication of *Camera Work*. The last issue was distributed to the 37 remaining subscribers.

Stieglitz, at this time, was certainly not drained of energy. He simply had decided to expend his energy on more private projects. While still a gallery director, he viewed the work of painter Georgia O'Keeffe. She had sent him watercolors from Texas, where she worked as a schoolteacher. His response when viewing them was, "finally, a woman on paper."[18] After a time, O'Keeffe ventured to New York, began a stirring relationship with Stieglitz, and ultimately married him.

When his public activities subsided in 1917, Stieglitz began to make portraits of O'Keeffe. Hundreds of exposures were made, many radically different from accepted contemporary practice in portraiture. Her hands, her feet, her clothes were often the subject of these images. Often Stieglitz omitted her face from the frame. Concurrently, he photographed strong portraits of friends and colleagues. When this body of work was exhibited at the Anderson Galleries in New York during the 1921 season, they created a tremendous response. Critics were barely able to describe the impact of the photographs. Some attributed the strength of the portraits to the strength of Stieglitz's personality, suggesting that he harbored some sort of hypnotic power.

In answer to these critics, Stieglitz turned his camera to the sky. He began a series of cloud photographs, calling them "Equivalents," as he believed them to be representational equivalents of human emotion. When this series was complete, he once again started to photograph tangible objects—though his themes were varied. One month he would photograph New York City from a window, the next he would focus on the landscape of his family home at Lake George, New York.

Poor health forced Alfred Stieglitz to cease active camera use in 1937. However, prior to this and in response to the entreaties of various artist friends, he started organizing exhibitions and ultimately opened two more galleries—the Intimate Gallery in 1925 and the American Place in 1929. To the end of his life, Stieglitz was a champion for the arts and its practitioners. When asked how much a photograph cost, Stieglitz would fire back, "How much are you willing to give the artist?"[19]

An American Place, located on a top floor corner space, which provided excellent lighting, was an active and vibrant mecca for both artists and collectors. During the 17 years that he held court at this last gallery, Stieglitz accepted the work of only two new artists—Ansel Adams and Eliot Porter. In 1946, Alfred Stieglitz was alone, in what was affectionately called "The Place," when he suffered a heart attack. He died a few days later.

From the time I was a child I never liked to do things according to the rules. I always made up my own rules, my own game....And I would feel in time the rules would...be changed.[20]
—Alfred Stieglitz
(1864–1946)

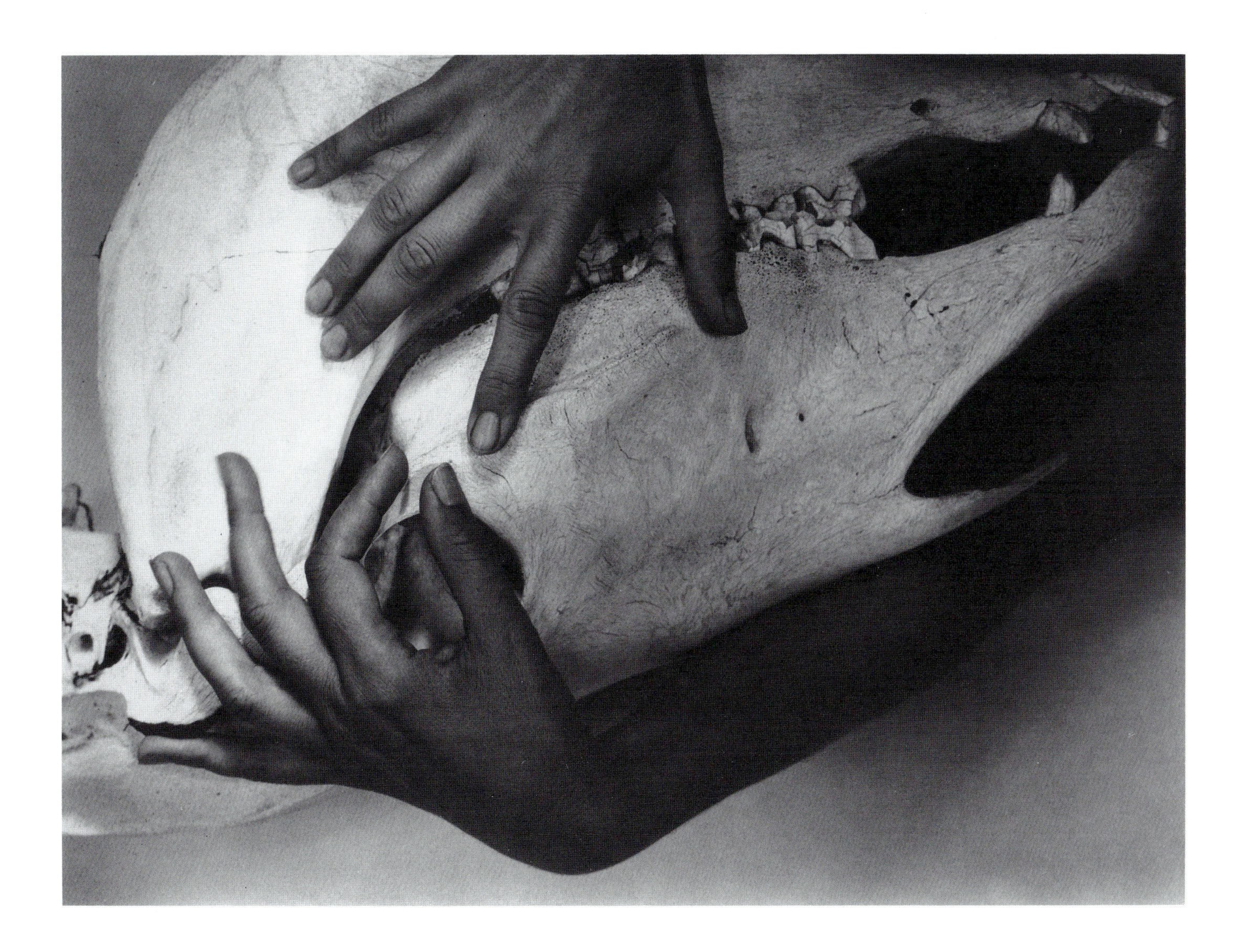

8-7
Alfred Stieglitz
Hands and Thimble
1920, gelatin silver print
IMP/GEH

9-1
Arthur Rothstein
Fleeing a Dust Storm;
Cimarron County, Oklahoma
1936, silver print
Library of Congress

9

The Farm Security Administration (FSA)

Was it journalism? Yes and no.
Was it history? Of course.
Was it education? Very much so.
If I had to sum it up, I'd say...we [the FSA photographic corps] succeeded in doing exactly what... we should do: *We introduced Americans to America.*[1]
—Roy Emerson Stryker

On October 28, 1929, the fabric of American economics unraveled. A combination of unchecked business growth, crazed financial speculation, and a decrease in investment funds, among other factors, caused a soaring United States stock market to plunge. That date signaled the beginning of what is now referred to as the Great Depression. The entire country was thrown into a panic. Businesses failed and unemployment skyrocketed. Banks repossessed homes and personal articles to repay delinquent loans. Business tycoons in tuxedos jumped out of skyscraper windows. Within a few years of the crash, over half of the Midwestern farmers saw their homes and acreage put on the auction block. The economy that had seemed so strong after the Allies' victory in World War I had disintegrated. The only entity powerful enough to remedy the situation was the U.S. Government.

Economic Politics of the Depression

At the time, the American government was headed by President Herbert Hoover, who spent much of his administration devising new economic and social programs to address the effects of economic disaster. For example, the government intervened in an attempt to solve the unemployment problem. The Reconstruction Finance Corporation was also formed. It was a federal program that made direct government loans to private businesses in an attempt to prevent further small business collapse. Hoover also tried to bolster the U.S. economy by advocating a 1-year moratorium on debts between international governments. The program eventually backfired, as 2 years after the moratorium ended, every country but one had defaulted.

The manner in which Hoover and the Republicans managed the economic slump and ensuing Depression made economics a natural issue for the next presidential election, held in 1932. Both the Republican and Democratic parties vied for votes using the Depression and ideas for economic relief as major issues. Hoover was the Republican nominee. Franklin Delano Roosevelt was the Democrats' choice. As many American voters blamed the incumbent Republican party for their appalling economic condition, Franklin Delano Roosevelt—with his "New Deal" policies—was elected a winner.

In his first inaugural address on March 4, 1933, President Roosevelt unveiled the New Deal. This series of programs was so named to suggest that each American could expect a "new deal" from his administration. The New Deal policies recognized a twofold problem—unemployment and debt. Many reforms that Roosevelt instituted to address those economic problems still exist today: the Federal Deposit Insurance Corporation (which insures and regulates the banking industry); the Securities and Exchange Commission (which oversees the workings of the stock market, among other organizations); and the Fair Labor Standards Act (which provides legislation regarding minimum wages and maximum work week).

Economic Rehabilitation Programs and the Formation of the FSA

Another program introduced as part of the New Deal, but which did not survive into the 1990s, was the Works Progress Administration (WPA). Started in 1935, its purpose was to create jobs, supervise all government relief programs, and transfer workers from relief rolls to government or private work projects. Many American dams, seawalls, and even sidewalks were constructed through this program. In fact, if you happen to be walking in an older U.S. community and notice a small metal plaque embedded in the sidewalk, it may denote that it was built as "A Project Funded by the Works Progress Administration."

The Resettlement Administration was yet another important program in Roosevelt's New Deal for all Americans. Its basic tenets were similar to those of the WPA, but they were specific to the problems of the rural population. Some of the objectives of the Resettlement Administration were low-interest loans to existent farmers, settlement of city dwellers onto communal farms for retraining skills, settlement of farmers into more suburban locations, and sponsorship of structured farm camps for migrant laborers. The Resettlement Administration was formed in 1935 and absorbed by the Department of Agriculture in 1937. When it shifted to the Department of Agriculture, it was given status as a separate agency with a new name: The Farm Security Administration (FSA).

Placing the FSA under the aegis of the Department of Agriculture was a wise choice, primarily due to the personalities involved. The new agency was headed by Rexford G. Tugwell, then Undersecretary of Agriculture. His was a powerful position, thus he was able to guard this fledgling agency against political machinations. While this arm of the New Deal was certainly not one of the largest or best funded, Tugwell was allowed his choice of advisory personnel.

Prior to his appointment to the FSA in Washington, Tugwell had been a professor of economics at Columbia University in New York City. When he assumed his new position, he called on a former student, Roy Emerson Stryker, to assist him in forming the Historical Section of the agency.

The main purpose of the Historical Section was to visually document not only the agency's activities, but the sociology of an America attempting to recover from economic crisis. Tugwell's choice of Stryker to establish policy and lead the Historical Section resulted in one of the most enduring photographic records of any era in United States history.

Roy Stryker and the Formation of the Photographic Corps

The road that lead Roy Stryker to Washington was not without complications. Born in Great Bend, Kansas, in 1893, Stryker was a member of a unique rural family. When not actively farming, Stryker's father spent his time reading classical literature and espousing radical political thought. Family discussions about economic and political theory were commonplace in the Stryker home. Having matured in this environment, it was natural that Roy Stryker, after World War I, left the farm for college. He chose to take classes at Columbia University in New York City.

Stryker, with his wife, found living accommodations in one of New York's many tenement houses. Their neighbors were mostly immigrants who worked feverishly simply to maintain the barest requirements of life. This experience impressed Stryker immensely. Food may have been lacking, clothing may have been scarce, but the fiber of these people was based on dignity, endurance, and the simple pleasures afforded by freedom.

Spurred by this exposure to poverty, Stryker decided to focus his studies at Columbia in the field of economics. It was there that he met a professor, later his mentor—Rexford Tugwell. Tugwell and Stryker agreed on many issues, and despite the fact that their relationship was occasionally stormy, they collaborated through the 1920s on a number of projects.

In 1924, Stryker was appointed assistant in Columbia's Department of Economics. He was an unorthodox professor. For example, he refused to employ textbooks. He growled that "You can't learn about labor problems from print. You have to become involved in the dynamics of the thing by actually being there. And that's what we did. We went to labor meetings. We went to printing plants, banks, produce markets, slaughter markets, museums, slums. I wanted the kids to see things for themselves."[2]

The students studied economic theory firsthand. They also learned to appreciate the power of photography. At this time, Lewis W. Hine (Chapter 4) and Jacob Riis (Chapter 7) were showing their imagery of the living and working conditions of the New York poor. When Stryker and his students saw these images, it was obvious that their impact was greater than that of the written word.

Stryker also appreciated the usefulness of photography for more ordinary documentation. As a farmer in the midst of the "bright boys at Columbia,"[3] Stryker was endlessly explaining the physical attributes of rag dolls, butter churns, corn testers, etc., to his teaching colleagues. He often found himself searching for photographs that would identify these objects. In desperation, Stryker started an ongoing list of ideas for photographs that would describe the simplest objects of rural life.

After he moved to Washington, Tugwell kept track of Stryker and his progress in the field of academics. Tugwell was impressed by what he had heard about Stryker and contacted him soon after arriving in Washington. Tugwell's request was clear. He wanted Stryker to help form—and eventually head—the Historical Section of the FSA. This photographically oriented project would ultimately generate over a quarter of a million images. Stryker jumped at the chance to, as Tugwell put it, "introduce Americans to America."

Stryker's job description (as written by Washington bureaucrats) looked, at best, boring. As far as they were concerned, he was to direct investigators, photographers, sociologists, and statisticians in their accumulation and compilation of reports, surveys, maps, and sketches. Stryker's idea of the job was quite different. Through his work at Columbia, Stryker understood the power of the photograph. He felt that photographic records would be far superior to graphs and maps as historic documents. Destitute migrants and prosperous farmers, eroded land and fertile land, human misery and human elation were all segments of the drama that Stryker wanted photographically portrayed. Quite naturally, the differences between Stryker's visions and those of the bureaucrats caused frequent rifts. Stryker was constantly criticized for not sticking to business (as the politicians saw it), for wasting taxpayers' money, and for stuffing a lot of silly pictures into government files. Time and again, Rex Tugwell, Stryker's boss, had to mediate between Stryker and the bureaucrats.

Assembling the FSA Photographic Corps

On receiving his assignment, Stryker's first order of business was to assemble a staff of talented photographers, for Stryker did not feel comfortable photographing. In his words, "I always had a camera but I had no more business with that damn Leica than with a B-29."[4] As an economist, he had no idea how to set up the darkroom. Stryker realized he needed help and sent for Arthur Rothstein, a chemistry major at Columbia who had a working, if somewhat scientific, familiarity with photography.

After arriving in Washington and arranging the project's darkroom and files, Rothstein struck out for the South and West to record the ravages of duststorms and droughts. He carried a Leica—a camera recently introduced to the United States (Chapter 2). It was perfect for the job, as it was lightweight, sturdy, and quick. Although Rothstein's photographic career had started by taking scientific pictures at a New York hospital, the drama of the Depression and widespread drought altered the scope of his concerns. His photographs for the FSA were far from surgically scientific. They were, in fact, very strong humanitarian documents.

Rothstein photographed with an intuitive sense of symbolism. For example, his *Fleeing a Dust Storm, Cimarron County, 1936* (Figure 9-1) highlights the farmer and his family, not as portrait subjects, but as allegories for all families devastated by the drought. Rothstein's sense of symbolism occasionally brought trouble to the FSA agency. The bleached skull (Figures 9-2 and 9-3) incident was a case in point.

While in the West photographing the "dustbowl,"* Rothstein spotted a steer's skull and horns lying on the ground. Thinking it a perfect symbol for the drought-plagued land, Rothstein took a picture. He then moved the skull a bit for compositional purposes and took another frame. In Roy Stryker's words, "That damn skull of Arthur's aroused more of an uproar than any other picture in history...Some editor out there in Dakota spotted the fact that Arthur had used different backgrounds and right away he labeled it a phony. Pretty soon front pages all over the country were saying the FSA was using phony pictures."[5] The controversy was eventually quelled when Stryker pointed out that while he moved the skull ten feet, he certainly did not move it out of the drought area, which was littered with such skulls.

Rothstein was the first of a handful of photographers ultimately hired to work on the FSA project. Some of the other photographers, such as Theo Jung, Carl Mydans, and John Collier, worked on the project for such a short time that their contributions were not markedly substantial.

*So named because the area suffered prolonged droughts and, subsequently, dust storms.

Others, such as Marion Post–Wolcott (one of the two women on the project), Jack Delano, and John Vachon, worked on the project for a more extended period. Each of these three had a definite style of shooting, producing in their images a particular "look." However, Ben Shahn, Walker Evans, Russell Lee, and Dorthea Lange are generally regarded as having been the strongest forces in the FSA Photographic Corps. Each of these image makers interpreted differently Stryker's premise for the project. Occasionally, the force of their imagery conversely altered Stryker's basic ideas as well.

9-2
Arthur Rothstein
Skull, Badlands, South Dakota
1936, gelatin silver print
Library of Congress

9-3
Arthur Rothstein
Skull,
Pennington County, South Dakota
1936, gelatin silver print
Library of Congress

9-4
Walker Evans
Atlanta, Georgia
1936, gelatin silver print
Library of Congress

Walker Evans

Walker Evans joined the FSA Photographic Corps in 1935, at the age of 32. His initiation to photography started eight years earlier, when he became fascinated by the visual and emotional content of postcard and newspaper photography. The exactness and the impersonality of this imagery for the masses struck a chord in the young Evans. He decided to follow this aesthetic, finding beauty in detachment and tension in ordinary composition. He demanded perfect texture and clarity, and therefore used a large–format view camera. His camera, which used 8 x 10–inch sheets of negative film, was bulky and cumbersome, but appropriate to Evans's unique vision. In 1934, an exhibition of his photographs of New England architecture was held at the New York Museum of Modern Art. The images hung for this show were quite stark, but incredibly rich in craft and composition.

Roy Stryker, who was always searching for competent photographers, hired Evans the year after his Museum of Modern Art exhibition. Of all the FSA photographers, Evans was the only one to use a large format camera. He was also the photographer least in sympathy with Stryker's social ideals for the project. While Evans photographed a great variety of subjects during his tenure at the FSA (everything from portraits to folk art objects), his style of vision was amazingly consistent. Many historians consider Evans to be a direct follower of Eugene Atget (Figure 10-5), others see his style as much more akin to Lewis Hine (Figures 4-3 and 4-4). In any event, his particular sense of nonemotional, surgical composition was markedly different than the style of almost all other imagery produced for the project.

After 18 months photographing in the southern portions of the country, Evans requested a leave from the FSA in order to work with author James Agee. The resultant book of photographs and text was titled *Let Us Now Praise Famous Men*. It took five years to complete and was the result of an extraordinary collaborative effort in the study of tenant farmers and their families.

Ben Shahn

Another interesting collaboration that had an impact on the FSA was that of Walker Evans and the painter–graphic artist Ben Shahn. In the 1920s, prior to the events that led to the formation of the FSA, Shahn and Evans shared an apartment in Greenwich Village, in New York City. Shahn had arrived in New York City from his native Lithuania to study at the Art Students' League. It was there that he met Evans. As they both shared a great interest in composition and form, albeit in different media, it was quite logical that their relationship prospered. In fact, it was Evans who instructed Shahn in the use of a camera, as Shahn had mentioned an interest in using photographs as studies for his paintings.

Both Shahn and Evans were hired by Stryker in 1935 (though at different times). Just as Walker Evans' aesthetic affected the direction of the Historical Section of the FSA, Ben Shahn's contribution was equally influential. He convinced Stryker that static photographs of lifeless objects were too limited in scope to truly capture conditions during the Depression. Photographs should dramatize these times, conditions, and people. Shahn suggested that rather than simply photograph an oft–used fiddle, an FSA photographer should include the fiddler as well. Shahn convinced Stryker of the importance of image vitality and a new photographic ethic was adopted.

Shahn would be the first to admit that his sensibilities were far from those of his friend Walker Evans. Evans normally used the large–format camera. Shahn used a 35–mm Leica occasionally fitted with a right-angle viewfinder (in order to catch his subject unawares). Evans was a master craftsman—all planes in his images were surgically sharp. Shahn often photographed so quickly, and in such poor light, that the resultant image was somewhat out of focus.

Shahn's evaluation of visual spontaneity over technique affected the other FSA photographers. When his colleagues in the project were bemoaning their own lack of technique, a missed opportunity, or shaky composition, Stryker would often point to Shahn. He would grab one of Shahn's images, shake it in the photographer's face, and remark that if Shahn could do it (make superbly moving images while not knowing one end of the camera from another), then anyone could.

9-5
Ben Shahn
Mountain Fiddler; Asheville, North Carolina
1937, gelatin silver print
Library of Congress

9-6
Russell Lee
Instruction at Home; Transylvania, Louisiana
1939, gelatin silver print
Library of Congress

Russell Lee

Ben Shahn was directly responsible for enlisting another talented colleague into the ranks of the FSA photographers—Russell Lee. Lee was important to the project not only for his empathic imagery of people in hard times, but also because of his unique relationship with Roy Stryker. As Stryker puts it, "There was some quality in Russell that helped me through my problems....When his photographs would come in, I always felt that Russell was saying, 'Now here's a fellow who is having a hard time but with a little help he's going to be all right.' And that's what gave me courage."[6]

Russell Lee was 33 years old when he heard about the photographic project going on in Washington. He checked with Shahn, a prior acquaintance. Shahn encouraged a meeting between Lee and Stryker, and although there was not an immediate slot for Lee, Stryker remembered him. When Carl Mydans left the project to join the staff of *Life*, Lee was hired to take his place.

Working for Stryker and the FSA, Russell Lee found his niche in the photographic world. His was the longest tenure of all photographers in the project. Also, his warm relationship with Stryker lasted far beyond the time when the Historic Section was phased into a different branch of the U. S. Government. It was with Lee that Stryker most often left the office to observe—firsthand—the society that his corps was capturing on film. As Stryker remembers, "One of my first trips out was to Minnesota...with Russell Lee. We were in a small town and he saw a little old lady with a little knot of hair on her head. He wanted to take her picture, but the woman said, 'What do you want to take my picture for?' Russell's response was part of my education as to how a photographer thinks. He turned to her and said, 'Lady, you're having such a hard time. We want to show them that you're a human being, a nice human being, but you're having troubles'."[7]

Russell Lee's photographic technique was at variance with, but sympathetic to, many of his colleagues. He was a humanist and always attempted to show the positive side of a situation. He was very interested in detail—occasionally spending a great deal of time photographing small objects that his subjects found particularly precious. When Lee found a subject of interest, he would often photograph it using a flashbulb. Unlike Walker Evans, Ben Shahn, or Dorthea Lange, Russell Lee often used artificial light to freeze his subject within the frame. Colleague Marion Post–Wolcott was the only other FSA photographer to use flash as extensively as Lee. Although the light provided by using a flash tended to visually "flatten" his subject, it provided the opportunity to shoot indoors or out. The use of flash further provided his images with a mesmerizing sense of detail. Not an iota of any portion of the frame went unilluminated. Although the lighting was flat and occasionally unflattering, it gave his work a firm sense of reality and honesty.

This directness and honesty was a reflection of Russell Lee's personal character. With his camera, he looked at situations squarely and he photographed them as such. Dorthea Lange called him "the great cataloguer."[8] It is disputable which of the two (Lange or Lee) cataloged their images more precisely—and which was more humanistic in approach to their subjects. It is certain, though, that Russell Lee had a far warmer relationship with his employer, Roy Stryker, than did Dorthea Lange.

9-7
Russell Lee
Lumberjack in Saloon; Craigville, Montana
1937, gelatin silver print
Library of Congress

Dorthea Lange

Dorthea Lange joined the ranks of the FSA photographers in 1935. In her five years with the group, she produced a number of startling, passionate images. Her tenacity in photographing and her great love for those photographs contributed to her role as one of the most memorable photographers in the FSA project. However, those same qualities also led Stryker to dismiss her from the project in 1940.

Lange's personality traits developed at a very early age. In 1902, she was stricken with polio. Humiliated by a withered right leg, she had to cope with childhood acquaintances who referred to her as "Limpy." In 1907, when she was 12, her father walked out of the house, never to return. He left his wife, Dorthea, and her brother Martin to survive in a small apartment in Hoboken, New Jersey. Dorthea's mother scrambled to find employment and took a job as a librarian in New York City—enrolling Dorthea in a nearby public school.

Thus, mother and daughter ferried from New Jersey to Manhattan five days a week. The neighborhood in which Lange's school was located had a profound impact on her future development. The school was near Hester Street, one of the most densely populated areas in America. It was crowded with immigrants who had recently arrived from Europe. As Lange remembers, "I was the only Gentile among 3,000 Jews."[9] It was certainly not a disparaging remark, as Lange loved the sense of ethnic richness inherent in that particular neighborhood. This exposure to other cultures served her well in later life, as it inculcated in her a sense of tolerance, indeed respect, for life-styles that differed from her own.

Other "foreign" situations were not as pleasant, but later proved themselves useful. After working at the library for a number of years, Lange's mother took a new job as a court investigator. As Lange recalled, "I remember my mother...making night interviews...in all kinds of wretched old Polish tenements in the winter.... I used to like to go with her."[10] Lange would wait on the corner for her mother, perhaps under a streetlight. However, it was a city and the neighborhoods were somewhat dangerous at night, so Lange mentally fashioned an invisible garment for herself. She called it her "cloak of invisibility," believing it would render her unseen—thus keeping her safe. She used this "cloak" often in later years, when wishing to be unobtrusive while photographing.

In her late teens, Lange enrolled at a school that trained teachers; however, she was not destined for a classroom. She knew, even before she owned a camera, that she wanted her life to include photography. To this end, she finagled apprenticeships with almost any photographer who would hire her. Perhaps the most important of these was Arnold Genthe (Chapter 4). Her only formal training consisted of one course with Clarence White (Chapter 10) at Columbia University.

Soon after the end of her apprenticeships, in 1918, Lange rebelled against conformity (and her mother's wishes). She announced that she and a friend were going to travel around the world. It was unheard of, at the time, for two unescorted women to travel much more than a few miles from their own homes. But she did go, ending up in San Francisco, California. As her funds were low, she took what was intended to be a temporary job as a retail clerk at a photofinishing company. One of the first patrons she served was a very kindly older man, Roi Partridge, husband of prominent photographer Imogen Cunningham (Chapter 10). A friendship developed among them, and through Roi and Imogen, Dorthea met many of San Fransicso's prominent cultural figures. Her new situation in California proved so comfortable that she decided to stay, and she subsequently opened her own portrait studio on Sutter Street. Soon she was attracting many wealthy and prominent clients. One of these was Maynard Dixon, whom Lange married in 1920.

Dixon was a well-known Western painter who specialized in canvases of Arizona and New Mexico. Soon after their nuptials, Lange and Dixon traveled in search of subject matter for Dixon's canvases. Lange took her camera along. While visitng an Arizona Indian reservation, Lange noted the effects of abject poverty on the Indian children. In response, she immediately began a series of noteworthy, compassionate portraits of these children.

When they returned to San Francisco, the country was beginning the financial breakdown that would ultimately result in the Great Depression. The economic situation particularly affected Dixon's artwork, as people no longer felt they had the money to spend on decorative artwork. The stress of financial concerns, among other factors, affected the relationship between Lange and her husband. In 1930 and 1931, in an attempt to save their marriage, they again ventured West, this time to New Mexico. It was there that they first noticed the jobless and homeless families who were moving their scant belongings from the East in search of some form of economic stability.

It is doubtful that most of these homeless migrants found stability. Neither did Lange and Dixon. Their relationship continued to be one of estrangement after their return to San Fransisco. However, the impact of having seen these migrant people affected Lange deeply. One day in 1932, as she was looking out the window of her studio, she saw a scene that would go far in directing her photographic future.

On that day, while puttering in her studio, she chanced to look out the window and saw a group of men huddled around the doorway of a local mission, or breadline. This charitable entity had been underwritten by a wealthy, philanthropic San Francisco lady, whom people called the "White Angel." These hopeless men reminded Lange of similar people she had encountered while traveling with Dixon. She responded by grabbing her camera, and, accompanied by her brother, she hurried from her studio to photograph the mass of men on the street. The image that she ultimately chose to print (Figure 9-8) was one that succinctly summarized current conditions for many American citizens. Soon, working in her studio was secondary to photographing the street people who were struggling to survive in the midst of economic calamity.

Lange began to exhibit these compassionate pictures of her fellow citizens. After one such exhibition, she met the noted economist

9-8
Dorthea Lange
White Angel Breadline; San Francisco, California
1933, gelatin silver print
©1990 The Oakland Museum, Dorthea Lange Collection

Paul Taylor. They shared many sympathies, as Taylor was interested in the plight of migrant workers, as well as the effects of drought on the economy of the South and Midwest. Taylor was a social scientist who compiled facts, but he also wanted evidence—in the form of photographs—to stress the conditions on which he was reporting. Their collaboration was a natural affiliation. Their first project together, on self–help cooperatives, evolved into something greater than simply a working relationship. By 1935, both Lange and Taylor divorced their previous spouses and were married.

This union was powerfully directed. Lange and Taylor worked on a number of government projects together. One of these, in 1935, involved working for Rexford Tugwell and Roy Stryker. Lange's position as photographer did not start auspiciously. Her first correspondence with Stryker was a request for funds to purchase equipment and supplies. Stryker responded by citing severe federal fiscal constraints.

As became her habit, Lange circumvented the system by paying for her expenses out–of–pocket, then requesting reimbursement. This stubbornness was but one of her character traits that exasperated Stryker. Fortunately for the project and the American public, Lange was tenacious. In her first major trip, in the spring of 1936, she shot the image that Stryker himself conceded was, "the ultimate....To me, it was *the* picture of Farm Security."[11] Lange called it *Migrant Mother* (Figure 9-9).

Lange was driving her car early one evening after a long day shooting. Her eye caught a sign that read "Pea Pickers Camp." After some deliberation, she turned down the drive and was confronted by a forlorn grouping of tents. One tent was simply a piece of cloth rigged to the chassis of a car that had no tires. Within this meager enclosure sat a group of children with their mother. The children were dirty—the mother exhausted. Lange spent less than 10 minutes with these people. In that time she learned that the pea crop had frozen and the family was living on scavenged vegetables and whatever birds the children had managed to kill. The woman was 32 years old and trapped. She had just sold the tires from the car to buy a bit of decent food.

9-9
Dorthea Lange
Migrant Mother; Nipomo, California
1936, gelatin silver print
Library of Congress

Migrant Mother epitomized Lange's chosen subject matter. In the words of filmmaker–writer Pare Lorentz: "You can usually spot any of her portraits because of the terrible reality of her people....They have the simple dignity of people who have leaned against the wind, and worked in the sun."[12] Lange took very detailed notes while photographing, understanding the historical context in which her photographs existed. In fact, the captions that accompanied each of her images were one of the few matters of accord between her and her boss, Roy Stryker. Although Stryker respected Lange's work, he could never convince her to act as a team player. As mentioned, Lange could not wait for government financing to photograph. She simply paid her own way and submitted receipts after her journey. This practice embarrassed and infuriated Stryker.

Another bone of contention between Lange and Stryker concerned her insistence that she control her imagery. Although Stryker respected her for her imagery (as well as her captions), he did not care for her evasiveness when the time arrived to send her negatives to the archives in Washington. She wanted to keep the negatives in her possession in order to oversee development, printing, cropping, and image selection. After finally acknowledging that she was not a technician, she hired Ansel Adams (Chapter 5), among other West Coast photographers, to do the required darkroom work. In doing so, she felt that she had found an acceptable compromise in the disagreement between herself and Stryker.

Stryker also compromised, temporarily returning to Lange some of her earlier negatives. However, her procrastination in giving them back to Stryker drove him to distraction. He recalled, at one point, that "We practically had to send the sheriff to get them back for the files."[13] When, during one such episode, Stryker heard that Lange had borrowed negatives in order to have prints made for a museum show, and that she was having them retouched as well—his patience was exhausted. The next year, in 1939, Lange was informed that she was fired—effective at the beginning of 1940.

Lange continued to photograph for the rest of her life. She worked in the Japanese internment camps with Ansel Adams, photographing those imprisoned during a period of nationalistic hysteria. Among other assignments, in later years, she traveled in Europe, Asia, and Latin America while photographing for *Life* and *Look* magazines. However, the haunting imagery she created for the FSA project is considered to be her most memorable achievement.

The Twilight of the FSA Historical Section

Between 1935 and 1943, more than 270,000 photographs were taken, indexed, and cataloged for the FSA. In 1942, the Historical Section of the FSA was transferred to the Office of War Information. At a later date, over three-fourths of the images were again transferred—this time to the Library of Congress, where 70,000 of these images are on file. The discrepancy between these figures—270,000 to 70,000—owes much to duplication of subject matter, as well as to Stryker's rather unfortunate habit of punching holes directly in negatives that he did not wish to use for the project.

In 1943, after the agency transfer, Stryker resigned from the Historical Section. He took with him 1,300 prints, either in contact sheet or small image form. He then took a job with Standard Oil, for whom he built a legendary collection of industrial photographs. In 1962, Stryker returned to the West—to the small community of Grand Junction, Colorado. It was there that he stored those 1,300 prints that, to him, epitomized the FSA project. In 1962, Edward Steichen (Chapter 8), then curator of the New York Museum of Modern Art, displayed *his* selection of Stryker's FSA images in the exhibition, "The Bitter Years." The work Steichen chose was antithetical to Stryker's idea of the FSA. Stryker referred to Steichen's choices as depicting "misery in its finest hour."[14] Stryker felt that the credo of human dignity, not frailty, should be the overriding message of the FSA legacy. To this end, in 1972, Stryker, in collaboration with Nancy Wood, chose approximately 200 images that he felt more appropriately represented the spirit of the times and the spirit of the program. His selection was printed, in book form, entitled *In This Proud Land*.

9-10
Dorthea Lange
Hayward, California
1942, gelatin silver print
National Archives

Roy Stryker died in 1976. Asked once whether there could ever be another project such as the FSA, Stryker responded,

When the water temperature was right, when the salts in the river were right, the salamanders came out of the water and pretty soon human beings were created.
Farm Security was one of those freaks,
one of those salamanders. It can't happen again.
But something new will happen.
Something different.
I wish to hell I could be around to do it.[15]
—Roy Emerson Stryker
1893–1976

10-1
Imogen Cunningham
Magnolia Blossom
1925, gelatin silver print

10

Form and Design

If you set up a camera and you do nothing but push the button, the camera will fill up a panel with information. At that point the camera is nothing but a vacuum cleaner picking up everything within range. There has to be a higher degree of selectivity.[1]
—Ray K. Metzker

Form and design are attributes relevant to any object. All objects are designed with an innate form. However, judgment regarding the merits of any design or form are extremely subjective. Decisions regarding "good" design alter through the ages—dependent upon cultural and idealistic factors.

In the earliest days of painting, perceptions concerning form and design were relatively unsophisticated. Artists represented simplistic and "absolute" concepts in their canvases. When delineating historic figures engaged in heroic acts, the artist would represent the "hero" as a flawless being. The "villain," if symbolized, would be equally romanticized as a person of unmitigated evil.

If architecture were the subject of the canvas, the painter would strive to represent the building in its most idealized form. The landscape surrounding the structure could, however, be treated with a modicum of imagination. Trees could be added or omitted. The sky, whether it be light and filled with fluffy clouds or more ominously portrayed, was normally delineated fancifully. These landscape details were treated interpretatively in accordance with the current aesthetic vogue. However, for many centuries, approved design ethic required that the "form" and perspective of the building be drawn realistically.

As early painters were aware of the demand for architectural precision, they often used the camera obscura or camera lucida (Chapter 1). The artist would place the respective tool in an appropriate location and then simply trace the reflected angles and forms. However, the image that resulted was only as exact as the skill of the artist using the instrument. The concept of perfect form was, therefore, relative.

When the daguerreotype was introduced in 1839, a "perfect" rendering was finally possible. The human hand had been dismissed from the actual creation of an image. Scientists were understandably pleased by the exactitude with which a camera or solar microscope could record minutiae. While some artists lauded the daguerreotype for its ability to capture detail, a sizable faction was concerned that the photographic image afforded too much information. Those critics worried that only a human hand could fabricate true "art." The camera, to them, was simply a machine.

Pictorial Design

Between the 1840s and the 1890s, most photographers, in an attempt to have their work accepted, sought to appease these critics by employing various techniques when making their photographic images. Prints that combined portions of different negatives appeared by the mid–1840s. Whether these photographs involved simply printing in a more dramatic sky (artifice) or using many negatives in the production of a print (combination printing; Figures 12-5 and 12-6), they blended the "machine image" with "human hand."

However, manual manipulation was not always involved in the creation of imprecise photographic imagery. Julia Margaret Cameron's portraits of the 1870s involved a simple soft–focus treatment in her attempt to modify "too much detail" (Figure 6-9). Peter Henry Emerson's work during the 1880s contained a similar attempt to minimize detail. Emerson believed in the Helmholtzian theory of vision—that, inherently, cameras provide more definition than the human eye when momentarily glancing at an object. He therefore suggested using softer, longer, focal length lenses. These, he trusted, would more successfully approximate natural vision, by obscuring unnecessary detail.

Cameron's portraits and Emerson's Helmholtzian theories were early examples of the quest to discover a distinctive photographic form or style. By the 1890s, groups of photographers had formed camera clubs to discuss new techniques and methods by which they could achieve artistic acceptance. They strove to have their photographs regarded in the same light as other, established, two–dimensional art forms by provoking emotion through their imagery. Photographs, they stated, should be regarded as highly as "pictures." Thus, they were referred to as pictorialists.

In their disdain for detail–filled snapshots, the pictorialists downplayed the fact that their images were made by machine. Stylistic individuality was highly regarded. For that reason, pictorialists usually chose to work with photographic materials that could be manipulated to suit the mood of the artist. Gum–bichromate, bromoil, and carbon were favorite media, and heavily textured paper was often a chosen support. These emulsions could be erased, etched, and otherwise manipulated to guarantee that no two images were identical. The textured paper obscured much detail, further removing the image from one that could be easily identified as originally having been generated by a camera. (Pictorialists are discussed in more detail in Chapter 8; their manipulative techniques are featured in Chapter 12).

Pictorialists usually chose idealized subject matter, whether it be landscape, portrait, or nude. They designed their imagery carefully, in most cases following compositional rules established by the painting community. Following current painterly design formulas, a sense of timelessness was evident in the work of the pictorialists. Demachy (Figure 12-7), F. Holland Day, Edward Steichen (Figure 8-5), Alfred Stieglitz (Figure 8-3), Frederick H. Evans, and Clarence H. White (Figure 8-1) were all at one period ardent pictorialists, striving to elevate the medium of photography from that of "handmaiden of art" to a distinguishable and unique art form.

The Search for Photographic Form

In Europe, near the beginning of the twentieth century, a group of photographers formed an alliance, which they named the Linked Ring. They were disenchanted with established photographic societies which promoted either "detail–filled" snapshots or camera images based on painterly composition. By 1902, Alfred Stieglitz had formed a similar group in the United States—the Society of the Photo–Secession. The aims of both groups were pictorially oriented. However, by 1907, Stieglitz began to revise his outlook on pictorialism. The softness of definition and idealized subject matter inherent in most pictorialist images seemed, to Stieglitz, to be overdone and hackneyed. Although Stieglitz juried the International Exhibition of Pictorial Photography in Buffalo, New York, in 1910, his aesthetic sensibilites were, by that time, more aligned to the new movement called modern art (Chapter 8). Some of his secessionist colleagues were not in sympathy with Stieglitz's change of heart. Original members of the Photo–Secession, led by Clarence White, left the fold in 1916 to form their own group—The Pictorial Photographers of America. In the late 1900s, other photographers chose to align themselves with Stieglitz's aesthetic and followed his lead in using the precepts of modern art in their photographic endeavors.

Alvin Langdon Coburn and Edward Steichen were two notable photo–secessionists who, by 1908, had abandoned the traditional impressionistic pictorial style in favor of exploration into a more radical search for form. Steichen often used a newspaper photographer's hand camera and photographed everyday subjects with a snapshooter's methodology. Coburn exhibited images taken in urban settings. Both were strongly geometric in their orientation and their images certainly were not intended to evoke passive reactions. In fact, in 1908, when Coburn's new work was exhibited, a fellow pictorialist complained bitterly to Stieglitz regarding Coburn's radical vision.

This criticism of Coburn's work went unheeded by much of the established art world. By 1913, Coburn's geometric designs within the photographic frame had garnered him solo exhibitions in Europe and America. After Coburn devised his vortoscope in 1917, he abandoned realism altogether. The vortoscope resulted from taping together three mirrors in the manner of a kaleidoscope. When photographing with this contraption, all sense of reality disappeared (Figure 3-5). In fact, when corresponding about his vortographs (images made with a vortoscope), Coburn remarked, "It does not really matter 'which way up' a good Vortograph is presented."[2]

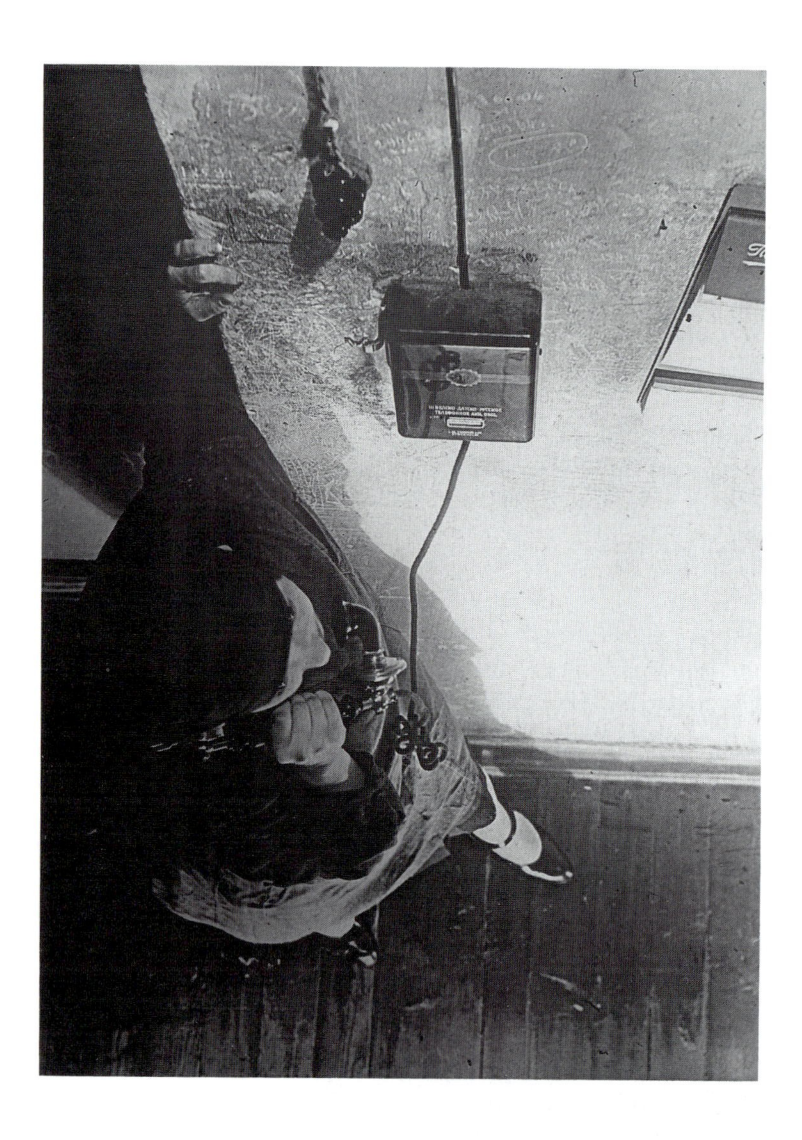

10-2
Alexander Rodchenko
At the Telephone
1928, gelatin silver print
MOMA

Further exploration into new, informal photographic techniques materialized quickly in the 1920s. Christian Schad, a member of the Zurich–based Dada group, updated William Henry Fox Talbot's photogenic drawing technique when he placed objects on photosensitive paper. Called "schadographs," these images were precursors to Man Ray's "rayographs" and Moholy–Nagy's "photograms" (Chapter 12; see also Figures 10-3 and 12-8). All were photographs made without cameras. All were completely abstract. All made strong statements about the changing parameters of photographic form and design.

This avant–garde aesthetic extended to Europeans and Americans who were eager to embrace new perspectives in photography. Sometimes these new perspectives were so radical that—in the words of Russian Constructivist El Lissitzky, when critiquing the publication, *America, Picture Book of an Architect*—"You have to hold the book over your head and twist it around to understand some of the photographs."[3] Photographers were abstracting even the most familiar subjects. Coburn, Alexander Rodchenko, Moholy–Nagy, Francis Bruguière, and Paul Outerbridge were photographing objects at different angles and with new sensibilities. In fact, Rodchenko refused to accept any photograph that had been taken from a normal vantage point. He disdainfully referred to such images as "belly button photographs"[4]—the angle of vision normally seen when an average person looks straight ahead.

Man Ray and Moholy–Nagy used many manipulative techniques in their search for "the new form." Solarization, multiple exposure, and negative prints were all fertile ground for their new vision (these techniques are explained in Chapter 12). When Moholy–Nagy left the German Bauhaus just prior to World War II, his avant–garde design aesthetic resurfaced at the Chicago Institute of Design. He influenced a new generation of photographers, among them Harry Callahan, in a search of an abstract, extended vision. In New York, Lottie Jacobi was combining straight imagery with light graphics and Barbara Morgan was experimenting with montage and cameraless imagery to create a new vocabulary of form in photography.

Montage was a popular method to extend photographic vision (Figures I-3 and 12-9). Photomontage, photocollage, and phototypography were visual vocabularies used by many artist–photographers during the 1920s and 1930s. This imagery, often torn from mass–produced periodicals, was recombined in varying fashions—often to express a political or cultural viewpoint.

New Ideals—Design for the Printed Page

Photographs had been regularly featured in periodicals since the invention of the halftone plate. Single images, centered on a page, functioned simply to illustrate accompanying text and were a standard design feature in most journals, newspapers, and magazines. In 1923, with

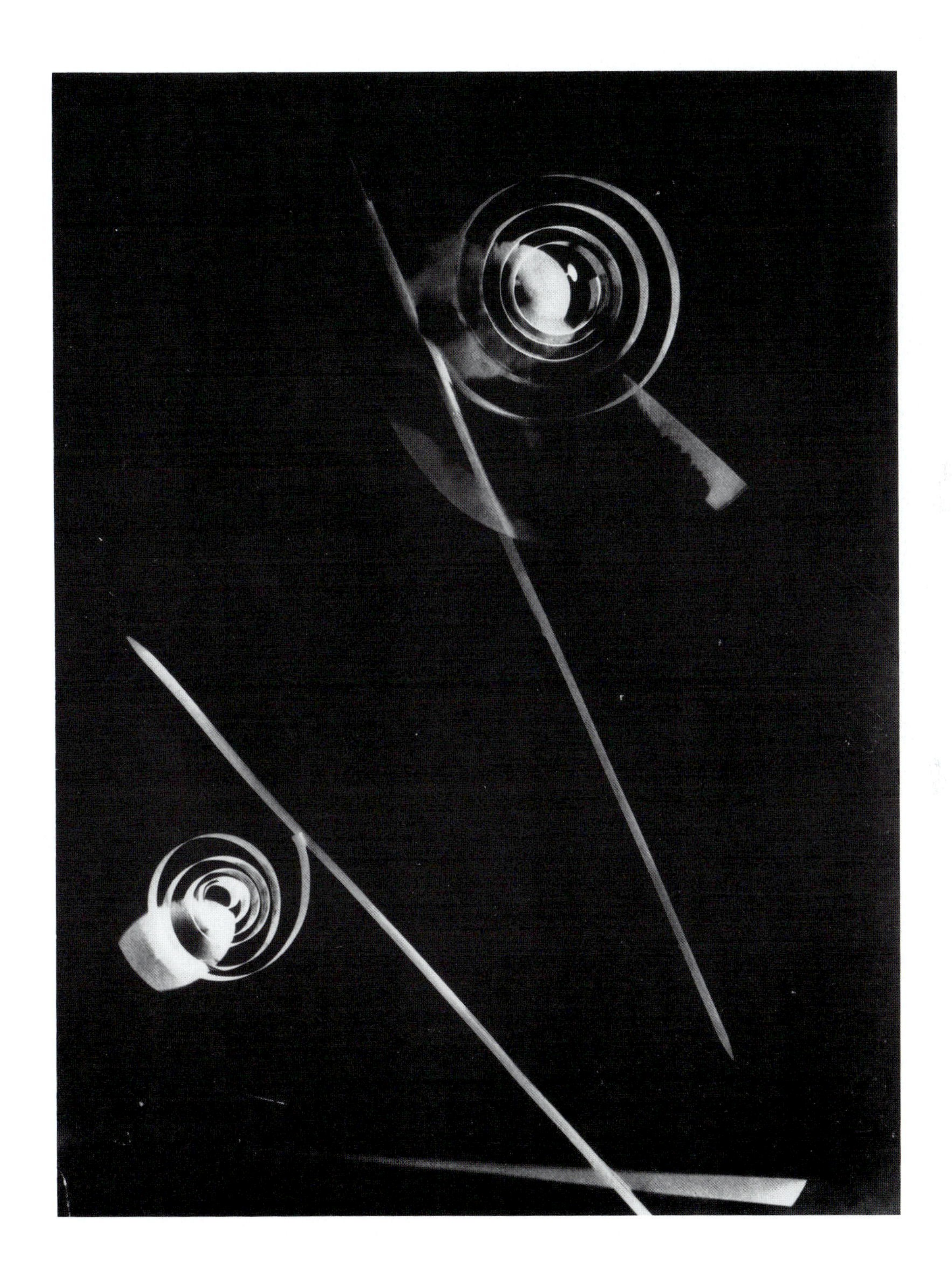

10-3
Laszlo Moholy–Nagy
Photogram
ca. 1925, gelatin silver print
IMP/GEH

10-4
Martin Munkacsi
Lucile Brokaw, Long Island, N.Y.
1933, gelatin silver print
courtesy of the
estate of Martin Munkacsi

the introduction of faster lenses and more portable cameras, photographic illustration became a bit more spontaneous. However, layouts remained visually stagnant until the introduction of 35–mm film improved the relationship between text and image. Picture editors, who formerly had only disconnected shots from which to work, now had a choice of 36 images on each roll of film—they were able to form relationships among images, not simply between photographs and text.

Beginning in 1928, these new dynamics in design layout were radically exploited by German editor Stefan Lorant of the *Munich Illustrated Press* (Figure 7-5). He experimented with new, but formal, juxtapositions between images. These images would be printed to span the alley, or space between two pages, and would then be enclosed by small black frames to isolate them on the page. Occasionally, Lorant demanded printing reversal, a technique of inking the entire background of a page black with the text printed in white.

Lorant was naturally drawn to hiring photojournalists who shared his daring new ideas of design. Martin Munkacsi, Felix H. Man, Alfred Eisenstadt, and André Kertész were just a few of the photographers hired by Lorant to create his picture essays. When the Nazis took power in Germany, Lorant fled. He settled in Great Britain, worked for the *Weekly Illustrated* for a while, then founded two of his own picture magazines—*Lilliput* (1937) and the *Picture Post* (1938). In 1940, Lorant moved to the United States to assist in the creation of Henry Luce's *Life* magazine.

Luce was obviously very influenced by Lorant's sense of design. He also approved of Lorant's choices regarding photographers. Felix Man and Alfred Eisenstadt were but two of a number of photographers who had originally made their names with Lorant and subsequently were hired by Luce. In fact, Eisenstadt was hired as one of the first staff photographers for *Life* magazine (Figure 7-7). His contribution of well–designed, illustrative images was instrumental to the magazine's success.

Other European photographers, notably Martin Munkacsi and André Kertész, started their photographic careers in Europe and continued them in the United States. Both Munkacsi and Kertész were from Budapest, Hungary. Both left Eastern Europe while still young to pursue magazine photography in Western Europe. Both ultimately left for the United States to pursue careers in fashion and photojournalism. However, what made these two photographers particularly important was their avant-garde approach to designing singular images for the printed page.

Martin Munkacsi was born in 1896. At the age of 18 he was working as the editor of a Budapest newspaper—writing, interviewing, and photographing. He left that position abruptly after an argument with his editor-in-chief and went to Berlin, arriving with no money, a very small reputation, and no job contacts. However, he talked his way into the offices of the Ullstein Verlag publication firm. When he left this impromptu meeting, he had a three-year photographic contract secured in his pocket.

Ullstein was the largest magazine publisher in Germany. It was also the pioneer of new ideals in photojournalism—including that of the photograph as a message-bearing medium. This mandate, evidenced best in Ullstein's *Berlin Illustrated Newspaper*, suited Munkacsi perfectly. In collaboration with his editor, Karl Korff, Munkacsi created fascinating new designs for individual and group photographs. When Hitler came to power in the early 1930s, both Korff and Munkacsi fled to the United States to continue their careers in magazine publication.

Carmel Snow, editor of the successful American magazine *Harper's Bazaar*, was familiar with the work of Munkacsi and initially commissioned him to produce a fashion layout. Rather than shoot the models in a stagnant studio setting, he took them to a snowy Long Island beach. The resulting images earned Munkacsi a permanent contract with *Harper's Bazaar*, as well as renown for his vital, dynamic style of photography.

f/64—The Formal View

New designs in publication blossomed in the 1930s. Visual innovations were attempted and new formats were devised. However, not all photographers of this decade were interested in the commercial aspects of fluid design in a magazine format. The tremendously fast, extremely portable cameras introduced in the 1920s had no place in their visions. Given a choice between instantaneous production of a photograph and absolute detail in all portions of a composition—many photographers chose the latter approach.

In San Francisco, in 1930, a group of image makers formed a loosely organized group known as "f/64." They ascribed to design principles that required careful composition through study of form. Using large-format cameras and small lens apertures (hence the name), these photographers dedicated themselves to the limited production of precise images, which represented an ideal in formal purity.

Ansel Adams and Edward Weston (Chapter 5) were two of the founding members of the f/64 organization. Willard Van Dyke and Imogen Cunningham were also members of the original group. All of these photographers were attracted to detail in form—whether it be naturally oriented or in celebration of contemporary technology. Each of these photographers is best remembered for a particular style or vision, although none was restricted to one subject matter. Adams is best remembered for his precise views of natural monumental panoramas. Weston is known, in part, for elevating such banal objects as rocks, shells, and peppers to the level of pure art and composition. Willard Van Dyke studied the architectural result of new technology in the 1930s, although he may be best recalled as a documentary filmmaker. Imogen Cunningham used the large-format camera and her personal sense of design to create lasting images of both natural and technological subjects.

Born in 1883, Imogen Cunningham started her photographic career after graduating from college as a chemistry major. Immediately upon leaving school, she took a job working for Edward Curtis, best remembered for his photographic work with Native American Indians (Chapter 6 and Figure 4-6). After two years of working in Curtis' studio, Imogen left for further study in Europe. When she returned, she opened a portrait studio, married, and raised three children. She married Roi Partridge, the man who initially befriended Dorthea Lange (Chapter 9). When she and her husband settled in Oakland, California, Imogen met Edward Weston.

Weston's influence on Cunningham cannot be overestimated. When Cunningham began her photographic career, she used a soft-focus lens manufactured by Pinkerman Smith. This apparatus produced soft-focus pictorial imagery. When f/64 was formed, Weston decreed that any photograph not sharply focused in every detail, not contact printed, and not mounted on a plain white backing should be considered "impure." Cunningham put her Pinkerman Smith aside and adhered to Weston's strict principles of photographic detail in her formal compositions (Figure 10-1).

During this time she, as well as most other members of the f/64 group, used a large-format camera. The ability to manipulate the lens in relation to the film plane allowed for maximum sharpness, regardless of the subject. Also, large negatives obviated enlargement. Each silver crystal on the negative was printed to reflect its innate size. Thus, minimal detail was lost between the actual shooting and the resultant print.

Design in 35-mm Photography—1930s to 1960s

While members of f/64 (among others) used large-format cameras exclusively, another group of photographers was eager for any improvements on previous camera design which would make the machines smaller and more portable. Between 1910 and 1940, a number of cameras were produced to answer these dictates. Around 1910, the Model 1A Graflex and the Linhof and Speed Graphic 4 x 5-inch reflex cameras were introduced. These machines allowed greater latitude of composition in addition to increased portability. While some photographers simply reveled in the freedom of these cameras (and cared little about image design), others, such as Eugene Atget (between 1910 and 1925), used these cameras for their inherent ability to easily design a frame.

Atget referred to his street images of Paris as "mere documents." However, Atget's training as a painter gave him the ability to instinctively design a frame in which formal considerations were paramount. The 4 x 5-inch camera, even with a tripod, was considerably more portable than previous photographic mechanisms. This allowed Atget the freedom to visually capture areas of Paris that would have been less accessible using bulkier equipment.

In 1925, the small and infinitely portable Leica camera was introduced. It was another breakthrough in possibilities for photographic design. As mentioned earlier, the 35-mm continuous-roll format allowed for tremendous changes in magazine layout design. André Kertész, together with Martin Munkacsi and Alfred Eisenstadt, used the 35-mm camera to produce innovative designs in their commercial photographic assignments.

André Kertész, like Munkacsi, began his career in Hungary and subsequently left to pursue his vision in Europe. Like Munkacsi, he worked for a variety of European magazines, including the *Berlin Illustrated Newspaper.* He, too, emigrated to the United States in 1930, eventually securing a contract with, first, Keystone, then Condé Nast—a giant in fashion and design photography. As successful as Kertész was in his chosen commerical field of fashion, it was his sense of composition and form that remains his most powerful contribution to the medium of photography.

When Kertész first used a Leica, it seemed that it had been developed for his specific vision. Both his professional and personal photographs had a certain sense of simplicity and freedom, which were greatly enhanced by the lightweight, quick-shooting Leica. His images of people amidst their surroundings, as well as his more formal architectural studies, were all based on the ephemeral moment which sought to capture

10-5
Eugene Atget
Boulevard de Strasbourg
ca. 1925, gelatin silver print
Museum of Art, RISD

the never-to-be-repeated instant. His style influenced such contemporaries as Brassaï, a man drawn to the nightlife of Paris, but who had never photographed until Kertész leant him a camera. Kertész's sense of instantaneous design was also mirrored by the French photographer Henri Cartier-Bresson. Kertész and Cartier-Bresson shared many similarities. Both were Europeans who eventually emigrated to the United States. Both considered themselves photojournalists (in fact, Cartier-Bresson was one of the founding members of the MAGNUM photo agency; Chapter 7). Both were staunchly humanistic in their picture-taking ideals, although neither emphasized "famous" people or "historic" events. Each used Leica cameras. Kertész's speed when taking a picture was later known as "candid photography." Cartier-Bresson's instantaneous image making was described as "decisive moment" photography. In his book, *The Decisive Moment*, Cartier-Bresson explains the design aesthetic of 35-mm photography: "there is a new...product of the instantaneous...movement of the subject. Inside the movement there is one moment at which the elements in motion are in balance. Photography must seize upon this moment and hold immobile the equilibrium of it....You'll observe that [when reviewing a previously exposed photograph] if the shutter was released at the decisive moment, you have instinctively fixed a geometric pattern without which the photograph would have been both formless and lifeless."[5]

Robert Frank, in the 1950s, ascribed to Cartier-Bresson's philosophy, although he criticized the imagery. Frank thought that Cartier-Bresson was simply a photographic "composer," not one who "felt" his imagery. An émigré from Switzerland, Robert Frank had very powerful opinions, and feelings, about the American culture. His work with a 35-mm camera resulted in the book, *The Americans,* a volume that altered formal photographic perceptions dramatically (Chapter 4). Although one critic noted that Frank "produced pictures that look as if a kid had taken them while eating a Popsicle and then had them developed and printed at the corner drugstore,"[6] other contemporaries lauded Frank for his photographic concerns. Even the design of *The Americans* was somewhat revolutionary. Following the layout of Walker Evans's *American Photographs,* Frank's book was uniquely designed. In an era when most books were laid out in a busy magazine-style format, *The Americans* was sparsely designed. The photographs were printed singularly and only on right-hand pages. The left-hand pages were reserved for pagination.

Robert Frank's 35-mm photographic design vision strongly influenced a new generation of photographers. Such photographers of the 1960s and 1970s as Lee Friedlander, Gary Winogrand, Bruce Davidson (Figure 6-13), and Roy DeCarava designed their frames tightly. All elements included in the frame were necessary to the final composition. Due to the speed normally inherent in shooting 35-mm film, these compositional decisions were spontaneous, and they signaled an evolving formal design ethic.

Controlled Design: Fabrications and Computer Assistance—1980 to 1990

Following the principles set forth by Alvin Langdon Coburn in the early 1900s and the work of the dadaists and other modern artists, some photographers in the 1970s altered their design ethic to the point that groups of their images have no basis in reality; in fact, the design *is* the photograph. Such image makers as Jan Groover and Barbara Kasten deliberately obscure references to "normal" picture planes and size considerations to produce pieces that are intended to be viewed simply in terms of their formal content. The edges of the frame—or where the picture ends—are as important as the content within the frame. Thus, although both artists use color and form to elicit emotion, there is no inherent message included in the frame.

Other contemporary image makers use twentieth century technology to form images that have a more emotive content. Robert Heinecken's "videograms" are examples of his continuing exposé of mass media and the values (or lack of same) that television has wrought upon our culture. Heinecken, discussed in Chapter 12, exposes still film in cameras that are trained at television monitors during commercially produced programming. He then designs murals or installations that illustrate his feelings concerning the ubiquity of mass media and its effect on the public consciousness.

It is interesting to note that although Heinecken has control of the moment at which he presses the shutter to produce an image, the design considerations of each frame have not been his. They have been arranged by an anonymous producer in a master control room. The impact of the design, in Heinecken's case, rests with the choices he makes when designing the final photographic product (Color Plate 7).

Similarly, Nancy Burson uses many mass–produced images when fabricating her computer–assisted composite photographs. She feeds snapshots or other public–domain images into a computer terminal. Once these images are recorded in the terminal, Burson may select any number of details when designing the final image. Hers could be considered a second–generational design ethic. To make the final image, Burson selects relevant details, makes a composite, and reshoots the frame from a television screen (Figure 6-18).

Computer–assisted design is currently altering the face of many two–dimensional art forms. Small pieces of preexisting photographs are combined in a computer, and a print—which need not have any basis in fact—evolves. Legal considerations of image ownership are thrown into question. Does the original photographer or the image designer have partial rights to composite imagery? It is a question currently debated from the halls of established museums to the desks of most art directors. It is not a question that can be easily answered. However the legal debate is resolved, we can trust that the computer–assisted imagery that proves to be historically important will be that with a strong sense of integrity in addition to a strong sense of form and design.

The charm of the photograph lies not in the object, but in...balanced relationships.[7]
—Laszlo Moholy–Nagy

10-6
Henri Cartier–Bresson
Place de'Europe, Paris
1932, gelatin silver print
©Cartier–Bresson, MAGNUM

11-1
Eddie Adams
Viet Cong Executed
February 1968, gelatin silver print
Associated Press/Wide World Photos

II

Photography and Combat

START A COLLECTION OF WAR PICTURES
Start now to collect these magnificent
illustrations of...churches in ruins, soldiers in trenches,
sacked villages, fleeing refugees....

—Advertisement in the *New York Times*
November 5, 1914

11-2
A newspaper sketch artist. During the Civil War (1860s), newspapers and periodicals employed artists to record scenes of battle. Although they occasionally embellished the skirmishes they observed, these artists provided the only mass-produced visual record of "live" action.

War, or armed conflict between groups of humans, has pervaded culture since man first populated the earth. Ancient cave paintings depict groups of tribesmen battling each other with crude weapons. The "artists" who made these drawings used tools as unsophisticated as the people and weaponry they portrayed.

Although the medium of photography was invented in 1839, there are several reasons why "sketchers" continued to depict war for the two decades after photography was invented. First, until the mid–1850s, exposure times were too long to capture any degree of action. Plates exposed for the required 3 to 20 seconds could show only the static aftermath of skirmishes. This was not completely acceptable to a public who was interested in vivid depictions of battle scenes. Their curiosity was somewhat satisfied through such magazines as *Leslie's Illustrated Newspaper* and *Harper's Weekly,* which printed woodcuts and engravings of battle action. Second, the equipment needed to make wet–plate (or collodion) photographs was very cumbersome. It was necessary for a photographer to pack several cameras; many heavy glass plates packed in wooden crates, which were themselves packed in shock–absorbing larger crates; chemistry; basins; and large tubs of distilled water.

11-3
Roger Fenton's photographic van. In 1855, Roger Fenton and Marcus Sparling left England for Russia to photographically record the Crimean War. They fabricated over 300 wet–plate images during this project.

Roger Fenton, Félice Beato, and the Crimean War—1855

These encumbrances did not deter English photographer Roger Fenton from heading for the Crimea early in 1855. Although he was not the first photographer to capture war–related imagery (a few daguerreotypes were made during the Mexican–American War in 1846), his effort was one of the most dedicated. Fenton packed 700 glass plates, five cameras, and much other necessary paraphernalia into his photographic van. Accompanied by his assistant, Marcus Sparling, Fenton journeyed to the Crimea, a peninsula in the southern section of Russia. The French and the British had seized a few seaport towns; the Soviets were naturally attempting to drive them out; and Fenton wanted to capture these actions. He had been hired and outfitted

11-4
Roger Fenton
The Valley of the Shadow of Death
1855, salted paper print
Science Museum, London

by an English publisher and had the blessing of the War Office as well as the personal recommendation of Prince Albert of England.

Fenton returned to England with over 300 exposed plates, as well as stories of anguish and discomfort. He had contracted cholera while on his mission—the living conditions were abysmal—and many of his collodion plates fell victim to the travails of war. Due to the heat during that summer of 1855, Fenton reported that the collodion emulsion sometimes dried (thus preventing proper exposure) before the image was recorded.

The collodion or wet–plate process also required up to 20 seconds of exposure time in most situations. This fact prevented Fenton from even attempting to photograph most troop actions. He had to be content with photographing portraits of the soldiers, as well as the battlefield—long after the action had ceased. Figure 11-4, entitled *Valley of the Shadow of Death*, shows the aftermath of battle. It is a static image, but it depicts Fenton's desire to evoke the reality of war through the inclusion of many cannon balls littering the landscape. In his Crimean imagery, Fenton most often chose to highlight the weaponry designed for aggression and death rather than the victims of battle—fallen warriors littering the field.

Other photographers at the time were not as squeamish about portraying corpses as was Fenton. One, Félice Beato, who first worked in the Crimea and then in India during the Sepoy Mutiny (1857), traveled alone to China to photograph the Franco–British expeditionary force. He was one of the first foreign photographers to enter the Chinese nation and was the first to dramatically portray the bloody carnage of war. Corpses abounded in his imagery. When they had not fallen in a compositionally appropriate spot, Beato instructed his coolies to reposition them. Thus, some of the same dead bodies appear in a series of different pictures.

The American Civil War

In 1861, the United States belied its name and divided into two factions—the Union and the Confederacy. Each armed against the other to provoke the most hideous of all conflicts—civil war. Conflicting loyalties pitted brother against brother, and father against son. When the American Civil War commenced, it was viewed by many as a passing skirmish. A photographer, Matthew Brady, was one of the first to recognize the historical importance of visually documenting this dispute. To this end, he left his two very lucrative portrait studios to hire photographers and outfit vans with equipment (Chapter 6).

It is evident that Brady understood photography's importance in history. While still a commercial portraitist, he published a book entitled *The Gallery of Illustrious Americans*. The

11-5
Félice Beato
Head Quarter Staff, Pehtang Fort August 1, 1860
1860, albumen print
MOMA

volume highlighted subjects who were instrumental in changing the course of American history; many wielded power in Washington. In part, these contacts with influential politicians paved the way for Brady to petition, and gain, immediate access to the frontlines of battle at the onset of the Civil War.

Brady could not have imagined the horrors and dangers of war when he and his equipment arrived at the Battle of Bull Run. In fact, his van of equipment was caught in shell fire, and he was nearly killed. He was lost for three days and eventually turned up in Washington, filthy and exhausted. However, this experience did not daunt his fervor for capturing the war on his collodion plates. He was soon back on the frontlines with his van of equipment, which the soldiers called the "what is it?" wagon. Brady was referred to as "that grand picture maker."

11-6
Timothy H. O'Sullivan
A Harvest of Death, Gettysburg, Pennsylvania
1863, albumen print from collodion negative
MOMA

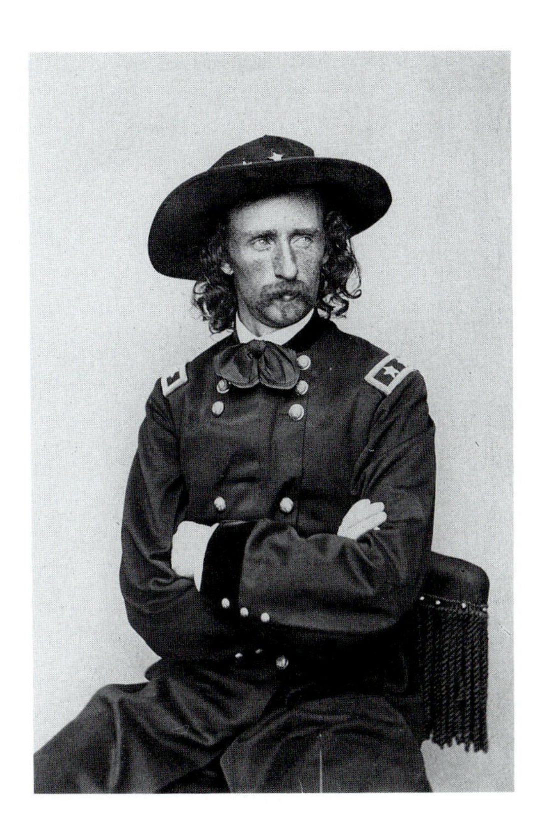

11-7
Matthew Brady
George Armstrong Custer
ca. 1864, albumen print
IMP/GEH

Fearing that he would not be able to singularly document all battles or occurrences relative to the war, Brady hired additional photographers, each equipped with individual wagons of photographic gear. Chief among these were Alexander Gardner (who had been in charge of Brady's Washington gallery) and Timothy O'Sullivan (who would later become famous for his surveys of the American West; see Chapter 5). These two, and hundreds of other picture makers, soon swarmed the fields where battles were in the making. They photographed every imaginable aspect of the war—battlefields (before and after skirmishes), soldiers, bivouac compounds, weaponry, officers, and corpses. When peace was declared, more than 7,000 negatives had been processed. Most of these bore the imprint of Brady's Photographic Corps and had been copyrighted by Brady.

These imprints and copyrights were not exactly honest representations of the work that had been completed. Although Brady *was* the employer and in charge of the corps, he copyrighted work performed by his employees on their free time—work that should have remained property of the individuals. This understandably caused a great deal of animosity, so it was no surprise when, in 1863, Alexander Gardner broke his liaison with Brady. O'Sullivan (as well as a few other photographers) left concurrently, for the same reason. Within two years Gardner published a two-volume *Photographic Sketch Book of the War*, containing 100 original photographs with accompanying text. Appropriately, each photograph was accompanied by the name of the photographer as well as, when applicable, the name of the printer. Including original photographs in a book was not customary. For reasons of cost and unsophisticated printing technology, photographs for publication were usually transformed into linecuts or woodcuts. Although the public appreciated the veracity of original photographs, the images were not usually readily accessible. Thus, Gardner's *Photographic Sketch Book of the War* generated a great deal of excitement upon publication.

Brady also displayed many of his photographic war images at his galleries. These were well received by the public. However, the reality inherent in photographic images produced a dramatic effect. An editorial was published in the *New York Times* announcing that war photographs were being shown at the Brady gallery, but cautioned that, "These pictures have a terrible distinctness. By the aid of the magnifying glass, the very features of the slain may be distinguished. We would scarce choose to be in the gallery, when one of the women bending over them should recognize a husband, son, or a brother in the still, lifeless lines of bodies..."[1]

Technology and Change—1880 to 1947

By the late 1880s, technological improvements had facilitated the jobs of both photographer and publisher. Combat photography had progressed greatly since 1862, when the *American Journal of Photography* stated, in reference to chronicling the Civil War, that "There will be little danger in the active duties, for the photographer must be beyond the smell of gunpowder or his chemicals will not work."[2] The conversion from wet-plate to dry-plate systems, better cameras, and the invention of the halftone process (Chapter 7) were among the many improvements that allowed war correspondents to carry out their assignments more efficiently.

By the start of World War I (1914), efficiency and speed were the watchwords of all photojournalists and publishers. The "graphic artist," or "sketcher," in the trenches had all but been forgotten. Thus, photography played an important role in that first global conflict; for example, aerial photography for reconnaissance purposes served a major function in many battles. When first used during World War I, aerial photographs of battlefields were no more than snapshots. However, soon thereafter, it was discovered that sequential photographs, taken from exact heights and at precise intervals, could yield tremendous amounts of tactical information. Figure 11-9 reveals trenches, roads, and troops. If a trained interpreter compared this image with photographs taken days earlier, he could glean an excellent notion of where the enemy would move next.

Aerial reconnaissance photographs taken during World War II are in use today in order to locate and defuse bombs dropped in Germany 45 years ago. Not all bombs exploded when dropped on the cities of Berlin, Hamburg, and Frankfurt. In 1983, almost 40 years after the conflict ended, a 1,000–pound bomb from World War II detonated near a school in West Berlin. Later, another was found under a nuclear reprocessing plant near Frankfurt. The importance of locating these unexploded munitions is obvious. Through use of old Royal Air Force reconnaissance photographs and sophisticated computer software, searchers have been able to uncover and defuse many of these literal time bombs.

11-8
Alexander Gardner
Home of a Rebel Sharpshooter, Gettysburg,* from Gardner's *Photographic Sketchbook of the War
1863, albumen print
MOMA

War Photography and Propaganda

The detail inherent in a photograph makes this medium a natural aid for reconnaissance work and historic documentation. Further, most people believe in the veracity of a photographic image; in the case of war photography, this trust is occasionally abused. As war is an emotional issue, many photographic correspondents attempt to skew public opinion through their selection of subject matter. Witnesses need not be impartial. Anyone or anything can have an inherent bias. When bias occurs in a photograph, it can become a political tool.

11-9
unknown photographer
Aerial Reconnaissance Photograph, Lavannes, World War I
1917, gelatin silver print
MOMA

For example, Roger Fenton's mission in the Crimea was politically motivated. The specific demand of his employer was to supply photographic documentation that would dispute newspaper reports that military leaders were grossly inefficient. During the Civil War, the Andersonville photographs, which showed skeletal, ill-clad Union prisoners, inflamed public opinion against the Confederate army. They were visually powerful documents, but they smacked of propaganda. Upon viewing the Andersonville images, Jefferson Davis—president of the Southern Confederacy—heartily disagreed with "the way in which they were taken and the manner in which they were used."[3]

Ever since publication of those Civil War images, photographs have been used to influence public opinion. Sometimes they are used to stimulate public thought and social reform (Chapters 4 and 7). At other times they are used to sway their viewers to divisively nationalistic thought. For example, Jimmy Hare's 1898 photographs of the wreckage of the *USS Maine* were described by writer Cecil Carnes: "He photographed swollen bodies with bones breaking through the skin; he took pictures of the emaciated living, and of babies ravaged by disease... Their influence upon public opinion can hardly be overestimated."[4]

Political and even commercial motivations occasionally prevent honest war reportage. This was evident in an exchange between newspaper baron William Randolph Hearst and American visual artist (although not a photographer) Frederick Remington. In 1898, Remington was in Cuba, purportedly to cover hostilities between the United States and Spain over control of that small island. After spending several weeks in Cuba, Remington was reported to have cabled Hearst with the message, "Everything is quiet. There is no trouble here. There will be no war. Wish to return." Hearst's reply was swift. "Please remain. You furnish the pictures and I'll furnish the war."[5]

As a publisher, Hearst was naturally, though selfishly, motivated by newspaper circulation. He understood that reporting any armed conflict resulted in increased circulation. Around 1900 (during the Spanish–American War) his *New York Journal* had a weekly circulation of 1.5 million, with as many as 40 editions issued in a single day. The United States readership demanded visual reportage on a continual basis. Hearst, with his newspapers, and publishers of the many illustrated magazines (such as *Leslie's Illustrated Newspaper* and *Collier's*), answered the public clamor for current news of the war.

Most of these images, though biased, were honest and unmanipulated. The same cannot be said for certain war photographs that were disseminated for particular, frequently divisive, results. During World War I (1914–1917), the German Kaiser Wilhelm I was dubbed "the Beast of Berlin" by many American and European publications. Photographs, posters, newspapers, even motion picture strips led one to believe that he made an active sport of cutting hands off of little babies.

In 1937, during hostilities that led to World War II, the Des Moines, Iowa, *Register and Tribune* published an exposé on war propaganda. This article showed, side by side, pictures released by the Allies and the Germans. The imagery provided by the Allies demonstrated German soldiers mutilating French and Belgian children. These were in direct contrast to the imagery made by German photographers, which showed these same soldiers treating the children with solicitude.

So, while photography had initially been regarded as the "mistress of truth,"[6] by 1914, following the publication of grossly faked propaganda, it was clear that "photographs could be tailored to any point of view and distorted to suit any camp or party."[7] The European dada movement recognized the shock value of reorganizing existing photographs into photomontage (Chapter 12). One member of the group, John Heartfield, used montage to produce highly dramatic political statements against the German Third Reich. Through the use of photomontage, Heartfield produced many posters, illustrations, and magazine covers lampooning the rise of Hitler, political trickery, and demagoguery. His most influential work involved the production of magazine covers for *Workers' Illustrated Newspaper (AIZ)*. The Nazis were so incensed by Heartfield's jibes at Hitler that they forced the Czechoslovakian government to remove the most offensive examples from a public display of Heartfield's work (Figure I-3).

From the early 1900s, the governments of Europe, the Soviet Union, and America were actively involved in using photography to manipulate public opinion. They understood the power of photography and its ability to alter national sentiment. Unfortunately, that realization led to photographic censorship during the First World War. Although hundreds of American photographers traveled to France, Belgium, and England to document the war in Europe, they were explicitly forbidden to visit battlefields. The Allied Command was fearful of public response to published imagery. To further ensure that no vivid imagery was made, the transport of cameras by correspondents was also forbidden. This did not deter the more resourceful photographers. Jimmy Hare, who earned the reputation of intrepidness during the Spanish–American War, was one of the first correspondents to arrive in Europe. He had ingeniously dismantled his camera and lens for travel, then had given the parts to a well–known publisher who smuggled them through customs.

In an attempt to further censor war news, the Allied Command attached a military officer to each correspondent. Their mission was to prevent the photographer or newsman from seeing or recording anything of significance. This method of censorship was so successful that early in the First World War, many of the larger publications simply abandoned any hope of continuous photographic coverage, and instead took whatever images they could secure and employed "sketch" artists to depict imaginary events for publication. The U.S. Government, among others, wanted to conceal a particular reality from the public.

In the Soviet Union during the 1920s, most types of photography were strictly censored by a government agency, the Bureau of Agitation and Propaganda. In Italy, during the mid–1920s, regional press associations were suppressed by the fledgling fascist government. Benito Mussolini, (as head of government), installed himself as the President of the Council of LUCE (acronym for the educational cinematographic union). The Ministry of Popular Culture fell under his direct supervision. The following memoranda were issued from that office: "26 August 1936: Do not print photographs in which the Duce [Mussolini] appears in the presence of monks" 6 April 1940: The authorities wish *Il Messagero* to print in tomorrow's edition two or three photographs of the Duce's visit to Nettunia. They particularly recommend printing photographs which show the artillery set up in firing positions."[8]

In Germany, during the period of the Third Reich, pictorial information was closely supervised by the Ministry of Propaganda. For example, all photographs of Hitler and Joseph Goebbels (Minister of Enlightenment and Culture; he has been cited in many history texts as the driving force behind the Holocaust) had to be submitted to officials before publication. The censors of public periodicals deleted any photographs showing German weakness, of Hitler wearing glasses, or of Goebbels' club foot.

American War Coverage and *Life* Magazine

In the 1930s, war photographers in the "free world" did not suffer such rigorous constraints. Certainly, military censors did attempt to edit some imagery; however, their restrictions were much less severe than those of the communists, fascists, and Teutons.

The Allied war correspondents were, in fact, enjoying unparalleled photographic freedom. First, they normally were not required to register themselves with any political agency other than the military when shooting in an active trouble spot. Second, they were enjoying the new flexibility inherent in the Leica 35–mm camera. Leicas (introduced in 1925) were easily portable, used long strips of gelatin roll film, which permitted repeated exposure, and had interchangeable lenses. Third, improvements in transportation greatly aided the war photographer in that it allowed these documentarians to arrive more quickly at a battle zone. Further, with the advent of the Wirephoto, they were able to transmit their imagery speedily to the many publications eager to present documentation of current events.

11-10
Margaret Bourke–White
Buchenwald
1945, gelatin silver print
***Life* Magazine**
©Time Warner, Inc.

Between 1936 and 1972, *Life* photographically presented wars in a most energetic and vivid manner. While not originally conceived as a war periodical, the first ten years of *Life* saw world conflicts erupting in all parts of the globe. China, Italy, France, the Soviet Union—all were actively involved in situations of strife. While *Life* used many photographs provided by picture agencies and freelancers, its major source of imagery was provided by its staff photographers—the original group of four at *Life* consisted of Margaret Bourke–White, Alfred Eisenstadt, Tom McAvoy, and Peter Stackpole. Each brought a particular talent to the pages of this wildly successful picture magazine. Eisenstadt had emigrated to the United States after working for avant–garde German illustrated magazines, bringing with him a cool polish to the concept of feature stories. Margaret Bourke–White was a former industrial photographer who was hired to work for *Fortune* magazine in 1929. Henry Luce, founder of both *Life* and *Fortune*, transferred her to *Life* upon its inception in 1936. Within two years, Bourke–White was assigned to cover events leading to the Second World War. (*Life*, Eisenstadt, and Bourke–White are discussed further in Chapter 7).

Bourke–White traveled worldwide. She was sent to Russia, North Africa (where she was the first woman granted permission to fly on a bombing mission), Germany—all areas of devastating conflict. When in Germany, at the close of the Second World War, she was one of the first photographers to enter the recently liberated concentration camps. Her arrival concurred with that of the liberating forces. This fact is shown plainly on the faces of concentration camp victims. She was clicking her shutter before the prisoners had time even to realize that freedom—not death—was imminent.

At the close of the Second World War, Bourke–White covered the painful separation of India and Pakistan, as well as the drama of the Korean War. Her intrepid spirit shone brilliantly during the Korean conflict. Occasionally, she would be in the wilderness for months at a time, photographing Korean guerrilla fighters. Her success as a war correspondent was due partially to the humanistic nature of her images. Her photographs of Mohandas Gandhi, of the survivors of Buchenwald, of the soldiers fighting in Europe, of the Korean conflict, highlighted people and situations that many other correspondents overlooked.

Robert Capa and The MAGNUM Picture Agency

At this time, most American civilian photojournalists normally fell into one of four catagories—free–lance, agency, syndicate (sometimes free–lancers sold to agencies and syndicates), or publications staff. Free–lancers shot film, then sold resulting images to agencies or specific publications. They did not work on contract or on specific assignments. Agencies were basically clearinghouses for imagery that was brought in either by free–lancers or by one of their own full–time stringers. Syndicates were large conglomerates that produced and distributed photographs for simultaneous publication in any number of national or local newspapers and periodicals. Publications staff photographers worked solely for one journal.

One agency that circumvented this hierarchy was MAGNUM, founded in Paris in 1947 by Robert Capa, Henri Cartier–Bresson (Chapter 10), David Seymour (also known as Chim), and George Rodger. Capa, christened Andreas Friedmann, was born in Hungary in 1913. At the age of 18, he was accused of alleged communist sympathies and was forced to leave his homeland. He moved to Germany, where he enrolled at Berlin University and worked part–time as a darkroom assistant, then as a photographer for the German photographic agency, Dephot. The Nazis, at this time, were gaining a disturbing amount of political power, and Andreas, a Jew, decided it was prudent to move again—his time to Paris.

In Paris, working as a photographer, Andreas Friedmann chose a new identity. He was now to be known as Robert Capa—a rich and talented photographer. He chose Robert for his given name, in reference to Robert Taylor, then a romantic screen idol. His surname was derived from film director Frank Capra (maker of the movie *It's a Wonderful Life*, among others). Andreas and his very close friend Gerta Taro decided that this combination would be simple to remember, easy to pronounce, and easy to recognize. Thus, Andreas Friedmann began selling his photographs using his new pseudonym, Robert Capa.

In Paris, Capa shared a darkroom with Cartier–Bresson and Chim (later cofounders of MAGNUM). In 1936, Capa began his incomparable career as war correspondent. It was then that he traveled to Spain to photograph the Spanish Civil War. His image entitled *Death of a Loyalist Soldier* gained him instant recognition and commissions to photograph the war for well–established American and European publications.

His dictum, "If your pictures aren't good you aren't close enough," was tragically realized when his close friend Gerta (who was covering the conflict in Spain for a different publisher) was crushed to death by a tank on drill maneuvers. Although terribly saddened by the tragedy, Capa continued to cover the Spanish Civil War. From there he traveled to China to cover the Japanese invasion. By 1938, Capa was declared "The Greatest War Photographer in the World" by the English magazine *Picture Post*. He was only twenty-five years old.

During World War II, Capa traveled extensively on assignment for *Life* magazine, which had also commissioned him to China. The collaboration between *Life* and the U.S. Army–Navy photographic pool circumvented some of the logistical problems inherent in war correspondence, such as availability of photographic supplies and speed of image transmission to publishers. However, this collaboration also created problems. Censorship became a mandatory process prior to publication. Also, darkroom workers who developed the film for correspondents could not always be trusted. For example, when Capa was in the first beach assault team on D–Day (June 6, 1944), he photographed in a frenzy of activity. It is obvious from the resulting imagery that he pushed his way backwards and forwards through the troops on that foggy dawn in Normandy. When the negatives arrived at the London office, they were to be developed, then immediately transmitted to the publisher. However, an excited darkroom technician used an excessive amount of heat in order to speed drying of the negatives. The emulsion melted off 98 of the 106 images that Capa had risked his life to take.

After the Second World War, Capa wanted nothing more than to be an unemployed war photographer. However, that was not to be, and in 1948 Capa went to Israel to cover the struggles of that new nation. In 1954, he traveled to Indochina to photograph French combat troops in North Vietnam. It was there that he stepped on a land mine and was killed. He was 41 years old, had covered five wars, and left behind a legacy of valor—his own as well as that of the multitudes of soldiers he had photographed.

11-11
Robert Capa
Death of a Spanish Loyalist Soldier
1936, gelatin silver print
courtesy MAGNUM

11-12
Robert Capa
D–Day Landing
June 1944, gelatin silver print
courtesy MAGNUM

David Douglas Duncan and the Korean War

Capa's connection to his publisher was circuitous. He was hired by *Life*, and as a result, photographed under the auspices of the American military photographic pool. Other photographers took a more direct approach. Edward Steichen (Chapter 8) enlisted directly into the United States armed forces. In World War I he was appointed commander of the Photographic Division of Aerial Photography for the American Expeditionary Forces. Steichen was again commissioned during World War II to re–create a similar type of photographic unit for the Department of Naval Aviation. W. Eugene Smith, with Steichen's encouragement, tried to enlist in this special unit. Unfortunately, Smith's injuries sustained while covering previous conflicts, coupled with weak eyesight, forced the Navy to reject his application.

David Douglas Duncan was 25 years old when the Japanese attacked Pearl Harbor, precipitating America's involvement in World War II. He promptly enlisted in the Marine Corps as a combat photographer. Duncan carried a carbine, as well as a camera, and flew 28 combat missions over Okinawa. His imagery and his personal valor won him great acclaim, as well as the Purple Heart and the Distinguished Flying Cross. He was the photographer assigned to record the signing of the unconditional surrender of the Japanese Empire at Tokyo Bay.

However, it was during the Korean War, in 1950, that Duncan made his most memorable images. He marched with the Marines of his old outfit, the First Division, while on commission for *Life* magazine. He photographed the struggle, the endurance, the exhaustion on the individual faces of his subjects. As he says, "I wanted to show what war did to a man....I wanted to show the way men live, and die, when they know death is among them, and yet they will find the strength to crawl forward armed only with bayonets to stop the advance of men they have never seen, with whom they have no immediate quarrel, men who will kill them on sight if given first chance."[9] Duncan fulfilled his mission admirably.

The Vietnam Conflict and Other Undeclared Wars

The Korean War was unlike any war yet fought by American forces. It was termed an "undeclared war"—thus the soldiers fighting and dying were not deemed war heroes. Rather, they were part of what was considered a United Nations "police action." The Vietnam War followed this new genre of conflict. Indochina had been the scene of internal conflict for many years before active United States involvement in 1963. Robert Capa had been killed by a land mine there in 1954. Ten years later, in 1964, David Douglas Duncan, among other photographers, was on hand to fly to Indochina to record the conflict. Photographers were handled quite differently than in any previous war or police action. Gaining access to the war was simpler. If you were a bona fide journalist, you received a card that identified you as such. You were then on your own to travel and record any situation that you considered worthy of documentation.

There were many notable photographers who worked both as free–lancers for news services and for *Life* magazine during the Indochinese conflicts. Eddie Adams, Larry Burrows, Don McCullin, and Nick Ut are a few who are particularly remembered for images that will live in the public psyche forever. As Umberto Eco, the Italian writer, observed in his essay, "The Photograph," the "vicissitudes of our century have been summed up in a few exemplary photographs that have proved epoch–making. Each of these images has...surpassed the individual circumstance that produced it; it no longer speaks of that single character or of those characters, but expresses concepts."[10] Eco was, in part, referring to Eddie Adams's apocalyptic image (Figure 11-1). Nick Ut also provided a singular image which described the horror of war in an isolated frame (Figure 11-13).

Other photographers, such as Larry Burrows and Don McCullin, made images that described the Vietnam situation. Burrows was responsible for a powerful essay published in *Life* during 1965 entitled, "One Ride with Yankee Papa 13." In a series of 22 pictures, Burrows chronicled one of the many stressful and dangerous

11-13
Huynh Cong "Nick" Ut
Terror of War
1972, gelatin silver print
Associated Press/Wide World Photos

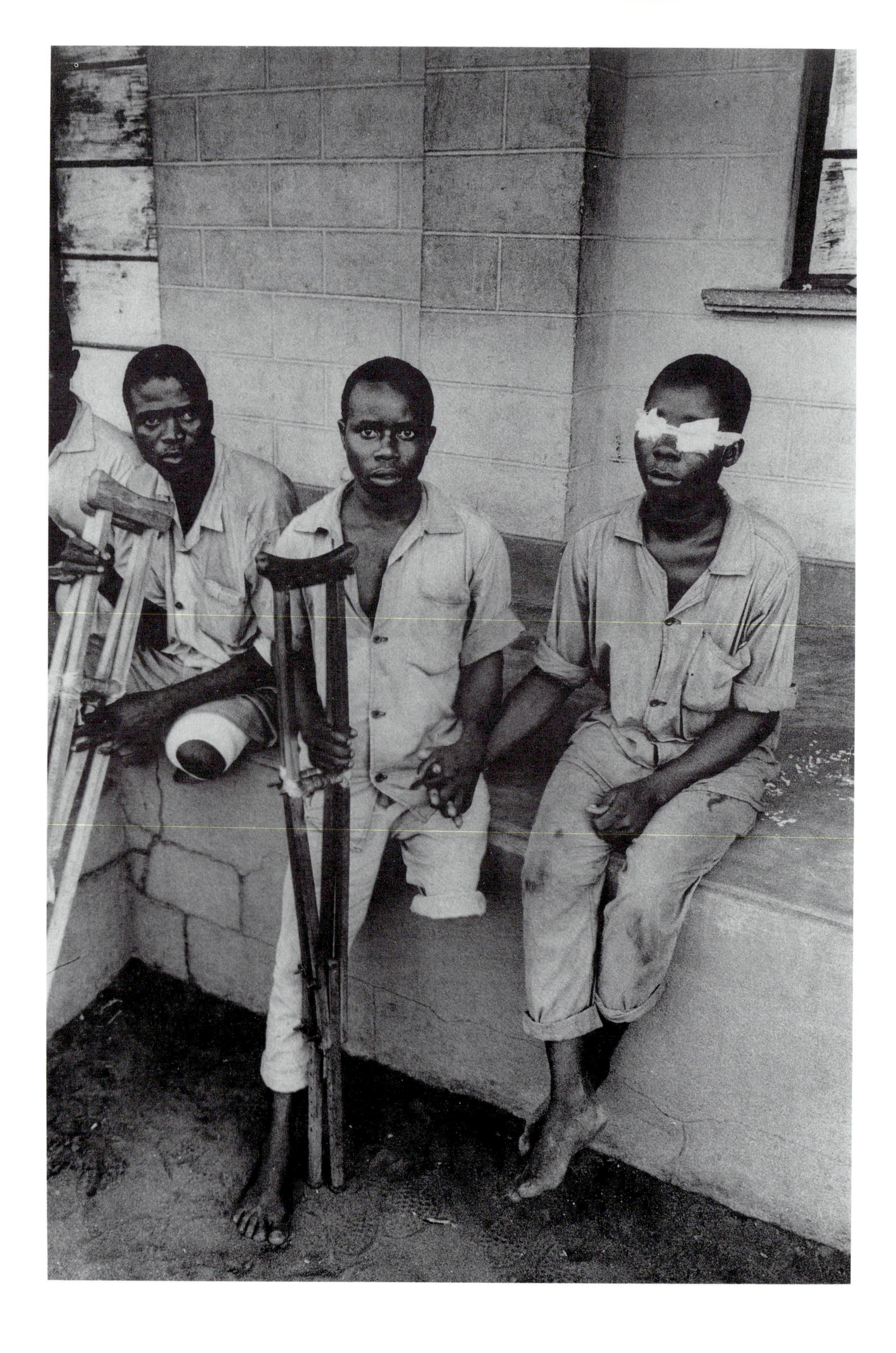

helicopter missions flown in Indochina. Readers of the magazine could not help but sympathize with the soldiers on this particular mission as they risked their lives to rescue a comrade who—as described at the end of the essay—died from wounds sustained in battle.

Don McCullin, an English photographer who had covered conflicts in the Congo, Biafra, Cyprus, and Ireland, as well as Vietnam, considers himself a "confrontational" photographer. He makes photographs that shock the viewer into confronting the situation of war, thus encouraging alternatives to armed combat. As McCullin says, "I have had an incredible taste of war and disgust for it....I'll always photograph the bad things. I look for the atrocity....I care, and I am going to photograph it—as horrible as it looks....The trouble is, here we are, and is anybody taking any notice?"[11] His photographic mission has been recognized and published in countless periodicals. His book, *Beirut: A City in Crisis*, chronicles his 5 years covering the urban civil war in that city.

McCullin, along with contemporary photographers Susan Meiselas, Gilles Peress, and others, continues to photograph ideological and political strife worldwide. Palestine, Iran, Afghanistan, China, Granada, Lebanon—the list unfortunately is growing, for although Robert Capa wanted nothing more than to be an "unemployed war photographer,"[12] reality dictates that employment for the brave souls mentioned here, as well as succeeding generations of war photographers, will be continued and steady.

War is a way of keeping down the population....War does provide a vent for the pent-up ferocity of man, and while man remains a savage, something of the sort seems to be needed....It has [also] been noted that man becomes restless or dejected after a generation of Peace....Peace seems to breed a disease, which, coming to a head in a sort of ulcer, bursts out into war....There is the suggestion which would probably be made by every other animal on the face of this earth, except man, namely that.war is an inestimable boon to creation as a whole, because it does offer some faint hope of exterminating the human race.[13]
—The Book of Merlyn
T. H. White

11-14
Donald McCullin
Biafra
ca. 1985, gelatin silver print
courtesy, ©MAGNUM

12-1
Jerry Uelsmann
Untitled
1975, gelatin silver print
Museum of Art, RISD

12

Manipulation

Any dodge, trick and conjuration of any kind is open to the photographers use....A great deal can be done and very beautiful pictures made, by a mixture of the real and the artificial in a picture.[1]
—Henry Peach Robinson
(1830–1901)

When one mentions a photographic image, one is usually referring to a factual scene that has been recorded on a piece of film in a fraction of a second. However, that viewpoint is unnecessarily limited. Almost since its inception, photography has not been restricted to split–second time frames, sophisticated optics, depiction of facts, or even the use of camera and film for image generation.

The first experiments with manipulating a photographic image transpired during the era of daguerreotypy (1839–1848). Soon after introduction of this first photographic process, hundreds of studios specializing in daguerreotype portraiture opened in the United States and Europe (Chapter 6).

Most of the portraits produced by these studios were very basic and relatively inexpensive black and white images. However, a few studio owners realized that there was room for stratification in that early world of portrait photography. Rather than keep their staff at the minimum number of personnel required to shoot and process photographic plates, they also hired unemployed painters to provide an additional feature. Billed by the business as "studio artists," these painters were, in fact, the first photographic manipulators.

Most often, the earliest manipulations were quite simple. After gilding (to strengthen the delicate surface emulsion), the artist would apply powered pigments to the cheek and lip area of the subject's image, making it appear more lifelike and, therefore, more visually pleasing (Color Plate 6). However (and for another additional fee), it was possible for the "artist" to *remove* small portions of the emulsion. If the sitter had an unsightly blemish, it would appear on the plate as "density." As the plate's density was very fragile, the painter would use a small tool—uncoated by any pigment—and simply brush away the offending fault.

One might have hoped that these underemployed painters would have confined themselves to brushing out—or applying—pigment to tiny areas on a photographic plate. However, by the late 1850s, certain unscrupulous painters used the photographic image as a literal "base" for their canvases. The ability to project photographic images onto paper and canvas allowed these "professionals" to almost completely obscure an original photographic image with heavy layers of oil color. The primary photograph was thus completely covered and rarely acknowledged as the basis for the work. The painter took entire credit for the remarkable verisimilitude of the canvases. It is no wonder that French author Charles Baudelaire contended that photography had become "the refuge of every would–be painter."[2]

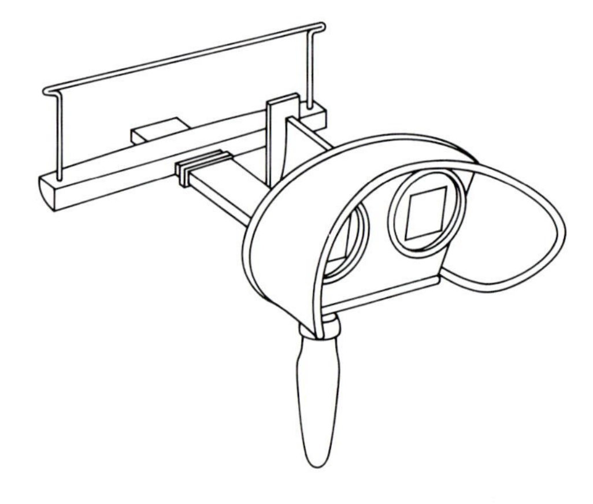

12-2
A stereoscope.
For 60 years, beginning in the 1850s, the stereoscope afforded its owner a "realistic three–dimensional" view of the world. Used with stereograph photographs, the system was both recreational and educational.

Early Photographic "Artifice"

Painters and photographers have, for over a century, enjoyed a lasting—if somewhat uneasy—alliance (Chapter 3). From the mid–1800s to the early 1900s, photographers recognized that only by emulating painterly form, technique, and compositional principles would their work be accepted by any faction of the "art" establishment. Toward this end, by the early 1850s, many photographers chose "scenics," or landscapes, as their subject matter. The reasoning was obvious, as renditions of landscapes were favored by the current painterly establishment.

The photographic process in vogue at that time was the calotype process. Photographic calotypists found landscapes to be relatively easy to render, as an immobile natural view could be effectively recorded during the necessarily long exposure time. However, calotypes had some limitations. They normally could not record the entire visual composition with the degree of detail demanded by current aesthetic dictates. If clouds were evident during the exposure, they would not be recorded, due to the length of exposure necessary for the "ground" area. In this case, the sky would appear a blank white (due to overexposure). If the clouds were to be effectively represented, the ground subject would lack detail (due to underexposure). The painterly principles of the day demanded rich and complete romantic detail throughout the entire image.

The problem was resolved during the late 1850s through what was then charmingly called "artifice." Numerous collodion plates would be exposed sequentially. Exposure times would be governed by the portion of the scene to be recorded. For example, the first exposure would be "normal," capturing the detail of the ground subject. The second would be a shorter exposure, which would highlight the sky and underexpose the dominant subject, the ground. The two negative plates, after development, would be printed using a "masking" technique. The "ground" negative would be exposed onto photographic paper while manually covering the area reserved for the sky. Then the negative containing details of the sky would be printed on the same sheet of paper, masking the ground area (Figures 5-4 and 5-5). The resulting "dual negative" image answered the demand for precise visual information throughout the image.

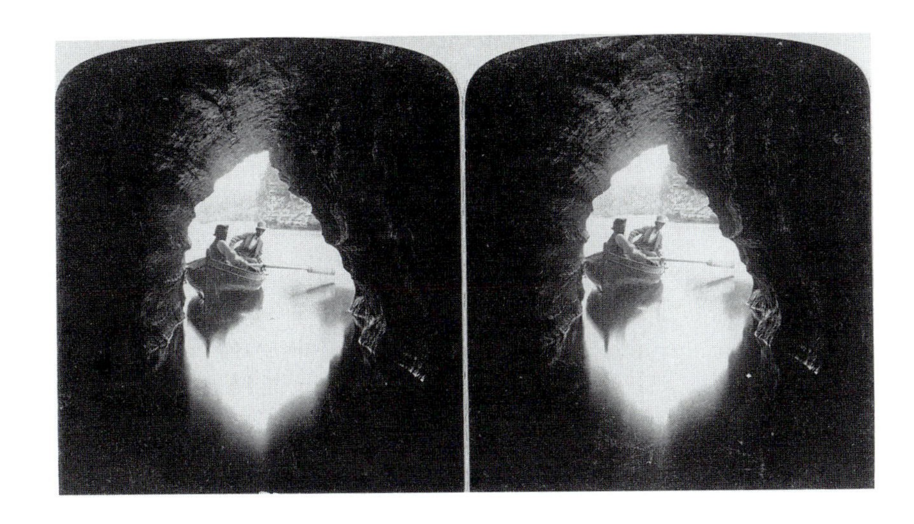

12-3
Henry Hamilton Bennett
In and About the Dells of the Wisconsin River
albumen stereograph
Museum of Art, RISD

12-4
unknown photographer
Her Guardian Angel
1898, stereograph
private collection

Stereography

Another manipulative technique that used landscape photography was the stereograph. Images were made using the stereo camera—a machine that took two simultaneous images of the same scene through two lenses placed side by side. The resulting plates were then printed and glued, again side by side, onto a firm cardboard support. When viewed through the lenses of a hand-held stereoscope (invented in 1861 by Oliver Wendell Holmes), the two images seemed to merge into a three-dimensional scene. Landscapes, soft romantic scenes using the human figure (occasionally as a wraith), and historic monuments were favored subject matter for the stereograph. By the mid-1850s, the collection and viewing of stereographs became a veritable mania. Stereoscopes were found as often in the drawing rooms of the 1850s and 1860s as television sets in the living rooms of today.

Combination and Collaged Imagery

By the mid–1800s, people were regarding the photograph as a form of amusement. Daguerreotypes and tintypes were produced in such numbers that they could be whimsically and frequently traded between friends. Stereographs were also affording hours of home entertainment by providing views of exotic places or equally exotic people. Images of popular society and cultural figures held a particular fascination. It was therefore unsurprising when relatively unmanipulated personal portraiture gave way to a new photographic venue, the collaged carte–de–visite.

During the 1860s, portrait studios began making composite photographs to advertise the range and quality of their products. These composites were created by cutting up individual portraits, gluing the many heads on a single sheet of board, then rephotographing the whole. The most successful shots (normally of beautiful, though unknown, women) figured prominently in the collage. However, soon after the original composites were introduced, an enterprising photographer realized potential profit in glimpses of the "rich and famous" through use of the photographic collage. Soon cartes–de–visite that collaged thematic photographs of historic or theatrical figures were on the market. They were instantly popular and were sold, or given away, as souvenirs at special events (Figure 6-6).

12-5
Oscar Gustave Rejlander
The Two Ways of Life
1857, combination albumen print
The Royal Photographic Society

The technique of collaging cartes–de–visite was clumsy at best. Relative subject size, aesthetic lighting techniques, and any sort of compositional ethic were rarely considered during fabrication. However, the manipulative technique of "combination printing," which was being explored simultaneously, allowed very precise composition and lighting.

The term "combination printing" denotes that the final single–surface print is made from a combination of a number of negatives. Although technically much more complex than the "dual–negative" landscapes, the impetus for making such images was the same. It stemmed from the continuing desire for photography to compete with painting as a fine art. During the Victorian era (namesake of Queen Victoria of England, 1819–1901), artists strove to represent idealized images. For that reason, still lifes were classical in their composition. Natural scenes were artistically arranged. Portraits and figure studies were carefully positioned. The Victorian belief that all art should uplift and instruct was followed carefully. Queen Victoria was a particular believer in the power of photography, as she saw in this medium the possibility of visually translating a cross–cultural reality that could promote her notion of an ideal society.

To this end, and prior to the advent of combination printing (during the early 1850s), human tableaux that exemplified a perfect social state were occasionally attempted. However, the cost of the models and the intricacy of shooting made these attempts almost futile. The flexibility inherent in combination printing allowed the photographer to shoot each model individually, then combine portions of each plate, after the fact, using a masking technique.

Oscar Gustave Rejlander and Henry Peach Robinson

Gustave Rejlander was the first to use this imaginative technique of combination printing. After experimenting with "dual–negative" landscape photographs, he encouraged an artificial (idealized) reality by using stage sets as backgrounds and actors as models. Occasionally, he would use portions of

12-6
Henry Peach Robinson
Fading Away
1858, combination albumen print
IMP/GEH

as many as 30 negatives to complete a final print. *Two Ways of Life*, an image he fabricated two years after his first attempt at combination printing, depicts the perfect Victorian ideal—the choice between good and evil, or between work and sloth. Queen Victoria was so pleased with the result that she purchased one of the five final versions. Unfortunately, the art critics of the day were not as appreciative, stating that "high art" could not be created through the use of "mechanical contrivances." Although Rejlander continued to create simpler combination prints, increasingly frequent barbs from contemporary art critics wounded him so deeply that wrote to his friend Henry Peach Robinson, "I am tired of photography for the public, particularly composite photos, for there can be no gain and there is no honor but cavil and misrepresentation."[3] Shortly thereafter, Rejlander bitterly retired from the public scene.

This was not the case with his correspondent, Henry Peach Robinson. Robinson also was assailed by critics, but he refused to be daunted by them. Initially a painter, Robinson used that experience in aesthetics to write, in 1869, a treatise that he hoped would elevate the status of photography. It was named *Pictorial Effect in Photography, Being Hints on Composition and Chiaroscuro for Photographers*. The artistic themes presented were based on painterly principles of composition. The book concluded with a chapter on combination printing. Robinson advertised his book as a "handy manual for the production of art photographs," and he illustrated the text with samples of his own work.

Robinson's ideas for combination images were first sketched on a piece of paper. He would then "fit" various negatives into this schematic with great precision. One of his most famous works, *Fading Away*, required the use of five negatives (Figure 12-6). Although the *Literary Gazette* panned the piece by stating, "Look steadily at it a minute...and all reality will 'fade away' as the make–up forces itself more and more on the attention."[4] Robinson continued to make manipulated imagery. He was a gregarious soul, convinced of his mission, who eventually wrote eleven books on aesthetics. His initial experience as a painter allowed his understanding of the aesthetic principles of photography as they related to those of the painterly establishment.

12-7
Robert Demachy
Struggle* from *Camera Work #5
1904, cliché–verre
UCLA Special Collections

Divorce from Reality—The Pictorialists

One reason that the established art world looked askance at the photographic medium was due to the camera—as they put it, a "machine" or "a mechanical contrivance." Cameras gave photographers the facility to endlessly reproduce images, mostly of dubious artistic merit. The only mark of distinction beween one print and the next was, perhaps, the signature of the artist. During the 1890s, the pictorialists (Chapter 8) countered this viewpoint through their novel use of photochemistry and choice of printing paper.

Bichromated gum emulsions and heavily textured printing paper were two tools that allowed these photographers the freedom to produce singular images that were never identical, either to the negative or to successive prints from that same negative. In addition, once having exposed the paper, the artist could manipulate the result through the use of fingers, brushes, and etching tools, in order to control highlights or to obscure details. These techniques produced monoprints and partially satisfied the criticism directed toward the camera as being simply a recording tool. Bichromates were hailed by collectors, who naturally preferred a "rare," or singular, and thus more valuable, image.

This "gum" or bichromate system of monoprint photography was first outlined in the book *Photo–Aquatint, or the Gum–Bichromate Process*. This publication was an invaluable sourcebook for photographers interested in singular pictorial representation. One of the authors, Robert Demachy, used bichromate regularly in his work. However, in the late 1800s, he also experimented with the cliché–verre process.

Cliché–verre was somewhat similar to bichromate in that it was a printmaking process. However, unlike a standard printmaking process, it involved the use of photographic paper. A cliché–verre print was made by covering a piece of flat glass with an opaque coating. An artist would then score lines through the coating, to the glass, using a small sharp tool. The resultant handmade "negative" would be placed in contact with a sheet of photosensitized paper. Occasionally, cliché–verre negatives were fabricated by using an existing camera–generated photographic negative and painting in, or scratching out, portions of that negative.

After enjoying a degree of popularity in the mid–1800s, the cliché–verre process fell into obscurity until it was resurrected by Pablo Picasso some 60 years later. Picasso had an interesting correspondent at the time, Man Ray—a person with whom he shared sympathetic viewpoints concerning altering established thought and aesthetics as they related to current art forms.

Man Ray

Born in America, Man Ray and his close associates, Marcel Duchamp and Francis Picabia, believed that society had become increasingly irrational and that artistic expression should translate that cultural modality (Chapter 3). Dubbed dadaists, they were interested in nonrepresentational imagery using any and all existent artistic media. For photographers, that meant that traditional boundaries concerning light, time, and cameras should be disregarded. They felt that all experimentation was valid in a search for new methods of visual interpretation and communication.

One of Man Ray's contributions to this new artistic methodology was discovered accidently. He was working in his darkroom when an unexposed sheet of photo paper dropped into the developing tray. He waited in vain for a couple of minutes for an image to appear. Nothing developed until he mechanically placed a small glass funnel, the darkroom graduate, and the thermometer in the tray on the wetted paper. He turned on the light, and before his eyes an image began to form. It was distorted and refracted by the glass more or less in contact with the paper and stood out against a black background. Man Ray dubbed the resulting image a "rayograph."

Man Ray continued to produce "rayographs" and in time, they became more complex as he experimented with light and motion during exposure. Occasionally, he incorporated the "antique" cliché–verre process in addition to movement and object placement. Strictly speaking, the entire rayograph technique was "antique" in that its basis was rooted in the experiments of William Henry Fox Talbot and his "photogenic drawings" (Chapter 2). The notion of resurrecting an old process did not affect or deter Laszlo Moholy–Nagy, who saw the rayographs in Paris and immediately decided to emulate them in his printmaking and photographic workshops at the German Bauhaus.* When Moholy–Nagy took the idea back to Germany, he dubbed the process "photogram," used it as a teaching tool, and he incorporated the process into his personal artistic work (Figure 10-3).

12-8
Man Ray
Rayograph
1922, gelatin silver print
MOMA

*The Bauhaus was the collaborative effort of artists and architects who wished to produce a new level of design and to provide an aesthetic for modern life. To this end, they formed a school, the "Bauhaus," in 1919. Headed by Walter Gropius, it was shut down by Hitler, as too radical an entity for Germany, just prior to the Second World War. A number of its adherents moved to Chicago, where they formed a "new" Bauhaus at the Institute of Design in 1937.

12-9
Laszlo Moholy–Nagy
The Law of Series
1925, gelatin silver print
MOMA

Rayograms, or photograms, exhibited the spacial and visual ambiguity that was vital to the dadaists. However, not all experimentation in the 1910s and 1920s involved cameraless images. Dada artists, and others, experimented in edge–reversal photographic techniques, solarization (in which a normally exposed print was re–exposed to light, altering its tonal values), and "negative" prints (in which the tonalities of a photograph were reversed). The subjects of these images were generally more literal, or "readable," than those evident in a rayogram, as the original negative had been generated by a camera. Still, these processes were a continuing effort to address the dada ethic.

Photomontage and the Collapse of Photographic Space

In Europe, during the early 1900s, a certain group of artists believed it important to divorce themselves from the "aesthetic elite" and to work on a more proletarian level. They were not part of the dada group, but some of their societal viewpoints were very similar. Through their work, they wanted to translate the incumbent spirit of society in a manner that could be easily understood by their fellow countrymen. The processes of collage and montage accurately reflected their times, those of "the chaos of war and revolution."

Moholy–Nagy, a vital force in the Bauhaus, was one of the first of this European group to experiment with photomontage. In using the photographic image, normally in conjunction with drawn geometric shapes, he presented a formal juxtaposition of symbols. In his work, the factual evidence that one expects from a photographic image would be countered with manipulation and spacial destruction. Moholy–Nagy's photomontages became such an important part of our artistic heritage that such late twentieth century painters as Robert Rauschenberg (Color Plate 5) and Andy Warhol have employed some of their spatial qualities in their imagery.

Warhol, primarily considered a painter, used the photographic image to examine certain aspects of the American culture (Chapter 3). Although his ideological concerns dealt primarily with the development of a new "pop" iconography in American art, his destruction of visual space closely mirrored the style of the Bauhaus, the Russian constructivists, and the dadaists.

Twentieth century photographers also used this "constructive" approach to alter spatial values and to provide a societal commentary. In the 1960s, Robert Heinecken, using the photographic–lithographic–silkscreen process, montaged imagery that spoke of the rampant sexism he believed to be inherent in American culture. Other contemporary photographers, such as Bea Nettles, manipulated spacial values by resurrecting the pictorial gum–bichromate system and applying three–dimensional objects to her imagery. Joyce Neimanas relies on construction of jigsaw–type imagery, which, when viewed as a whole, presents a readable image while altering a sense of realistic space (Color Plate 8).

12-10
Robert Heinecken
Invitation to Metamorphosis
1975, photographic emulsion on canvas
private collection

12-11
Bea Nettles
Flyin Princess Triptych
1978, bichromate print on vinyl

12-12
Ray K. Metzker
Untitled (beach scene)
n.d., gelatin silver print
Museum of Art, RISD

This technique is also true of the work of Ray Metzker, who, beginning in the 1960s, collapsed space by treating his film as one continuous plane of emulsion. The film was broken into frames when exposed through a normal camera lens. However, Metzker preplanned the relationship between these frames and actually used the black frame lines as part of his compositions, thereby warping our perceptions of spatial relationships. Another method Metzker used to generate visual excitement through collapsed space was collaged imagery. After shooting a roll of film of a particular subject, Metzker would develop the film, making multiple prints of certain frames, then collage all the prints to make a repetitve visual pattern. While the imagery shot was three–dimensional, the resultant piece stressed a two–dimensional orientation.

Manipulation of Exposure—Edgerton and Mili

Other late twentieth century manipulative techniques involved the use of shortened, extended, or multiple exposures. In the area of shortened exposure, the works of Harold Edgerton and Gjon Mili are notable.

In a sense, Edgerton and Mili were *not* manipulators, as they photographed exactly what was happening in an instant—through the use of stroboscopic light (Chapter 2). A stroboscopic light is a form of artificial "flash" illumination. When used, the light "strobes" or pulsates in milliseconds. If a camera were pointed at an object moving during this pulsation, each nuance would be "frozen" on film.

Edgerton and Mili worked together at the Massachusetts Institue of Technology. Edgerton's basis for experimentation was scientific. He photographed any number of objects that altered during movement, from the flight of a hummingbird to that of a bullet. Gjon Mili photographed in a mode more attuned to the media of the day. Hired by *Life* magazine, his work with the "strobe" highlighted subjects of popular interest.

Multiple Exposure

Other photographers manipulated photographic imagery through the use of multiple exposures on the same piece of film. This normally entails opening the shutter repeatedly while a single piece of film remains stationary in the camera.

Unlike today's cameras, photographic equipment in the early twentieth century was not equipped with a "locking" mechanism that automatically prevented multiple exposures. Therefore, the early camera operators had to remember to manually advance the film beween exposures or risk a visually confusing image. Some photographers, such as Edward Steichen, enjoyed creating controlled confusion through this type of camera manipulation. His multiple exposures of urban

12-13
Dr. Harold Edgerton
Back Dive
ca. 1954, gelatin silver print
©Estate of Harold Edgerton, courtesy of Palm Press, Inc.
David Winton Bell Gallery, Brown University

12-14
Harry Callahan
Alley, Chicago
n.d., gelatin silver print
Museum of Art, RISD

architecture in the early 1900s stand as landmark images of the genre. His work was recognized and emulated by Harry Callahan, who had met Steichen in the 1950s.

Although Callahan is known primarily as a "straight" photographer (Chapter 6), his background at Chicago's Institute of Design (the new Bauhaus) and his acquaintance with both Steichen and Moholy–Nagy affected his work profoundly. This is evident in Callahan's multiple exposure photographic imagery. The photographs were usually the result of chance. He would expose a roll of film (normally of a very limited and particular subject), partially rewind the film in the camera, then reshoot the film in order to re–expose the same emulsion. The effect created overlapping images—surrealistic and quite graphic.

Surreal Manipulated Visions

Surrealism is, according to Webster's Dictionary, a "modern movement in art…which purports to express subconsious mental activities through fantastic or incongruous imagery…"[5] This ethic has been of increasing interest to photographers in the last few decades. They have been aided in their quest for photographic surrealism by the fact that the technique—or science—of photography has been both expanded and simplified. By simplifying the methodology involved in the medium, artists are freer to explore a more personal, sometimes more manipulated, vision.

Some photographic tools that are widely available, and which can aid a personal manipulated vision, might be as simple as a wide–angle camera lens, as used in the work of Bill Brandt, Christian Vogt, or Emmet Gowin. Other photographers occasionally use plastic "toy" cameras as a device to explore a surreal vision that does not demand visual acuity.

Nineteenth Century Techniques, Twentieth Century Interpretations

Historical processes from the late 1800s to the early 1900s have recently been re–examined to aid in a new, manipulative vision. Henry Holmes Smith has re–explored the cliché–verre process. A student of Smith's, Jerry Uelsmann, has been making personal, surreal images for almost three decades (Figures 12-1 and 12-15). Uelsmann's work can be compared with that of Gustave Rejlander or Henry Peach Robinson. While many contemporary photographers blend a few negatives, Uelsmann has been known to fabricate a piece by using the numbers normally equated with a Rejlander. He repeatedly uses a personal iconography—or system of symbols—in his work. These symbols are used as communicative tools, much as Robinson and Rejlander employed trademarks of their day to complete a narrative thought. Uelsmann's work, however, gravitates toward an intensely personal emotion rather than the societal narratives that Robinson and Rejlander fabricated. Uelsmann is so convinced that the narratives produced by those two Victorian–era combination printers are valid, that his motto pledges that "Robinson and Rejlander live!"

A number of other artists are resurrecting nineteenth century manipulative techniques to communicate their twentieth century visions. In the early 1970s and 1980s, both Arnulf Rainer and William Parker made a numbers of portraits, then expressionistically overpainted directly on the photographic emulsion. During the 1970s and 1980s, Benno Friedman experimented with the photocollage and montage systems that were in vogue 50 years prior. Contemporary photographer Lucas Samaras montages a number of his images and occasionally produces a piece that hearkens back to the era of the carte–de–visite. Also, as mentioned, such photographers as Bea Nettles and Betty Hahn (another former student of Henry Holmes Smith) have interpreted their visions by working with the early 1900 gum–bichromate emulsion process.

12-15
Jerry Uelsmann
Untitled
n.d., gelatin silver print
Museum of Art, RISD

Breaking the Rules

Like all other art forms, photography is a continuum that bases current and future work on past techniques, processes, and aesthetics. Also, photography, like all traditional art–media, has always had an indisputable "author." During the 1920s, the dada ethic used imagery from popular publications in their photomontages, but in such a manner that the resulting image was transformed sufficiently to merit the signature of the montagist. During the 1960s, pop artists based their work on a celebration of a ubiquitous media. Cartoons, television imagery, even product packaging was translated into photograph, silkscreen, paint, and sculpture by contemporary artists. However, again, the resulting piece of art was far enough removed from the original to make "authorship" of the piece indisputable.

As we enter the twenty–first century, the extent of this media "appropriation" has altered. A number of photographers are currently utilizing popular imagery as if it were of their own making. They think of all imagery as part of a bank at which everyone has access to the same account. Richard Prince and Vicky Alexander pull images directly from media publications, paste them into close juxtaposition, and claim them as their own. Sherry Levine extends this practice to its ultimate degree. Levine rephotographs art made by historically recognized "masters," prints the image to precisely mirror the original (with the exception of size and surface), then signs them with her own name. Her titles are the only acknowledgement of prior authorship. For example, Levine rephotographed a Walker Evans image from the 1930s (Chapter 9), printed it without alteration, and signed the bottom, "Sherrie Levine, *After Walker Evans*."

In using most current color materials, photographers of the twentieth century are breaking yet another long–standing artistic pattern—that of archival quality. Archival, as it relates to art, means that the piece in question should remain stable and unchanged for at least a century after its fabrication. Most current color materials—be they color negatives or prints—simply cannot make that claim. Some photographers simply choose to disregard the question of archival quality, make their art using available methodologies, and hope for the best. Other current photographers, notably the Starn Twins, actively challenge the question of archival quality. In their early photographic constructions of the 1980s, they occasionally, but purposefully, use materials that they are assured are not archival. High–acid paper, inexpensively produced pieces of material, and cellophane tape are all used in their imagery. Their work is designed to be as transitory and changeable as our culture.

Fabricated to be Photographed

Still other current photographers actually construct tableaux to be photographed, then dismantled. Their vision has little to do with the nineteenth century ideal of "photograph as fact." Arthur Tress began in this venue during the 1960s by requiring that his human subjects juxtapose themselves with such eerie objects as dead rats and misshapen plants. More recently, he has constructed miniature stage sets, included seemingly unconnected objects, then woven in a literal meaning by titling his constructions.

Similarly, such artists as Joel–Peter Witkin, Eileen Cowin, and Nic Nicosia require that their human subjects position themselves within a preplanned area (occasionally set up as elaborately as a movie set) to present the dramatic vision of the photographer. Bruce Charlesworth and Cindy Sherman (Figure 6-17) share this methodology of photographing, but use themselves alone in the "set."

There has been an increasing movement, during the 1980s and into the 1990s, to use small doll figures in these fabricated tableaux. Obviously meant to symoblize humans, these dolls occasionally exemplify more than a simple artistic message. Ellen Brooks and Laurie Simmons use dolls and dollhouse sets to describe their viewpoint on the status of a gender–oriented society. David Levinthal cooperated with Gary Trudeau (of *Doonesbury* cartoon fame) in the compilation of a book entitled *Hitler Moves East: A Graphic Chronicle, 1941–43*. The book, published in 1977, uses small plastic soldier dolls. The irony of using children's toys in relation to combat is obvious. Levinthal's more recent work centers around large–scale color prints that idealize the American cowboy through use of small plastic figurines.

Sandy Skoglund often crafts her own dolls, and other life–sized objects in plastic resin. Multiple epoxy babies, cats, goldfish, etc., are then placed in preselected or fabricated "sets" and appear to be overrunning the area. These nightmare visions are accentuated through Skoglund's use of unnatural lighting and color. Occasionally, Skoglund will photograph friends in black and white, print the negatives in dramatic, unreal color, and photomontage various negatives to produce an unsettling vision.

Other "fabrication" photographers appropriate media imagery, print it relatively exactly, but apply verbiage to the piece in order to skew the original visual meaning. Robert Heinecken uses images from fashion catalogs (among other sources), reprints them without alteration, then typesets fantasy stories underneath the visual image. Barbara Kruger appropriates current visual media then integrates slogans directly into the photographic image. Hal Foster has written about Kruger (although his message is equally substantive when viewing the work of other "appropriators"), that they are "manipulators of signs more than makers of art objects...that renders the viewer an active reader of messages more than a contemplator of the esthetic."[6]

Computer–Based Manipulation

Now that the computer age has arrived, there is no way to know what methods will be employed in the future to manipulate our vision of the world, or in what physical manner photographs will be manipulated. The possibilities are endless. Computers make our notion of light, as a prerequisite for generation of imagery, obsolete. The boundaries beween art forms and their physical requirements are blurring. However, as long as the work produced by these machines produces a personal, communicative vision, we should be able to evaluate it as art. If not, we shall find ourselves back at the time of the pictorialists, who attempted to convince the world that machine–generated imagery is a valid form of artistic communication. It is important to remember that communication is the overriding issue when reflecting on any art form. While photography was initially hailed as a scientific discovery, then as a recording device, it was through manipulation of this medium that photography came to be regarded with respect in the art world. Photojournalistic documents communicate through the absence of manipulation. A more personal type of communication demands a manipulated photograph—one which sees beyond the boundaries of the "real" and into the realm of the "surreal."

I am interested in...reality as an illusion....
Seeing is not believing.[7]
—Duane Michaels

Color Plate 5
Robert Rauschenberg
Homage to Frederick Kiesler
1966, lithograph
MOMA

Color Plate 6
photographer unknown
Untitled, (Mother and Son)
19th century, daguerreotype
Museum of Art, RISD

Color Plate 7
Robert Heinecken
from the series *Docudrama*
(Maria Shriver and Steve Baskerville)
1984, cibachrome color print
private collection

Color Plate 8
Joyce Neimanas
Untitled #3
1980, photo–collage
Center for Creative Photography

1 From Photo–Graphy to Photography: 500 B.C. to 1840 A.D.

1. Reilly, Rose. "Steichen the Living Legend." *Popular Photography* (March, 1938).
2. *Collier's Encyclopaedia*, P.F. Collier and Sons, 1956, volume 4, p. 457, 4th edition.
3. Davy, Sir Humphrey. "An Account of a Method of Making Profiles by the Agency of Light," *Journals of the Royal Institution*, 1(1802): 170-174.
4. Foque, *La Vérité sur l'invention de la photographie: Niépce sa vie, ses essais, ses travaux*—Paris: Librarie des Auteurs et de l'Académie des Bibliophiles, 1867, p. 222.
5. Fox Talbot, W.H. "Some Account of the Art of Photogenic Drawing, or, the Process by Which Natural Objects May be Made to Delineate Themselves Without the Aid of the Artist's Pencil." R. and J.E. Taylor, London, 1839.
6. Ibid.
7. "Cômpte–rendu des Séances de l'Académie des Sciences." 8 (1839): 171.
8. Duca, Lo. *Bayard*. New York: Prisma Press, 1943, pp. 22-23.
9. Holmes, Oliver Wendell. "The Stereoscope and the Stereograph," *The Atlantic Monthly* 3 (1859): 738-748.

2 Photography: Inventions, Problems, and Solutions—1840s to 1990s

1. Hill, David Octavius. Quoted in *The Hill/Adamson Albums*, 1973.
2. *American Journal of Photography* 2 (1861): 42.
3. *American Journal of Photography* 6 (1863): 144.
4. Oute, Marc [pseudo]. "Gelatine." *The British Journal Photographic Almanac* (1881): 213.
5. Vogel, Hermann Wilhelm. *Philadelphia Photographer* (1873).
6. *Humphrey's Journal of Photography* 16 (1865): 315-316.
7. Herscher, Georges. "Conversations with Jacques Henri Lartigue," In *The Autochromes of Jacques Henri Lartigue*. New York. 1980.
8. Szarkowski, John. *Looking at Photographs*. New York: Museum of Modern Art, 1973, p. 114.
9. Turner, Peter. *Reading Photographs*. New York: Pantheon Press, 1977, pp. 83-86.

3 Camera and Canvas—The Alliance Between Painters and Photographers

1. Shaw, George Bernard. *The Amateur Photographer* (October 9, 1902).
2. "Upon Photography in an Artistic View, and its Relation to the Arts." *Journal of the Photographic Society* 1 (1853) 6-7.
3. Dwight, M.A. *Introduction to the Study of Art*. London, 1856.
4. Baudelaire, Charles. *The Mirror of Art*. London: Phaidon Press Ltd., 1955, pp. 228-231.
5. Escholier, Raymond. *Delacroix–Paris*. Vol. 3. Paris: H. Floury, 1929, p. 201.
6. Hawes, Josiah Johnson. "Stray Leaves from the Diary of the Oldest Professional Photographer in the World." *Photo Era XVI*, 1906, pp. 104-107.
7. Rosenblum, Naomi. *A World History of Photography*. New York: Abbeville Press, 1984, p. 249.
8. Zola, Emile. *Photo–Miniature* 21 (1900): 396.
9. Pennell, Joseph. *British Journal of Photography* 38 (1891): 677.
10. Ibid.
11. Frith, William. *The Studio* 1 (1893): 6.
12. *Photographic Notes* 8 (1863): 16.
13. Stieglitz, Alfred. "Exhibition Notes." *Camera Work* 12 (1905): 59.
14. Stieglitz, Alfred. "The Editor's Page." *Camera Work* 18 (1907): 37.
15. Stieglitz, Alfred. Interview. *Evening Sun*, 23 April 1912.
16. Hausmann, Raoul. Quoted by Jean Keim. "Photomontage after WWI," In *100 Years of Photo History*, edited by VanDeren Coke. Albuquerque, NM, 1976, p. 86.
17. *Art of the 20th Century*, Düsseldorf: Kubst Bierber, 1969.
18. Emerson, P.H. "Photography, A Pictorial Art." *The Amateur Photographer* 3 (1886): 138-139.

4 Social Documents—Photography and Social Culture

1. Brassaï. Interview by Paul Hill and Thomas Cooper. *Dialogue with Photography*. New York: Farrar, Straus, and Giroux, 1974, p. 40.
2. Thomson, John. *Street Life in London*. London: Bernard Blom Inc., 1969 (facsimile edition).
3. Jay, Bill. *Customs and Faces, Photographs by Sir Benjamin Stone, 1838–1914*. London, 1972.
4. Brassaï. Interview by Paul Hill and Thomas Cooper. *Dialogue with Photography*. New York: Farrar, Straus, and Giroux, 1974, p. 37.
5. Avedon, Richard. *Diary of a Century*, February 15, 1970, New York: Viking Press.
6. Ibid.
7. Stryker, Roy and Wood, Nancy. *In This Proud Land*. New York: New York Graphic Society, 1973, p. 9.
8. Kerouac, Jack. *The Americans*. New York: Aperture, 1969, p. ii.
9. Szarkowski, John. *The Americans*. New York: Aperture, 1969, cover notes.
10. Lyons, Danny. *The Bikeriders*, MacMillan and Co., 1968, foreword.
11. Lyons, Danny. *Conversations with the Dead*. Philadelphia: Holt Rinehart and Winston, 1971.
12. *Diane Arbus: A Monograph of Seventeen Photographs*. Los Angeles, 1980.
13. *Diane Arbus*, New York: an Aperture Monograph, 1972.
14. Kirstein, Lincoln. *American Photographs*. New York, Museum of Modern Art, 1938.

5 Landscape and Nature

1. White, Minor. *Rites and Passages*. New York: Aperture, 1978.
2. Dellenbaugh, Frederick. *A Canyon Voyage*. New York: G.P. Putnam and Sons, 1908, p. 58.
3. Ibid., p. 179.
4. *Harper's New Monthly Magazine* 39 (1869): 465-75.
5. Jackson, William Henry. *Time Exposure*. New York: Cooper Union Publishers, 1971, p. 205.
6. Jussim, Estelle. *Landscape as Photograph*. New Haven: Yale University Press, 1985, p. 21.
7. Newton, Sir William. *Journal of Photographic Society of London* 1 (1853): 6.
8. Muir, John and Adams, Ansel. *Yosemite and the Sierra Nevada*. Boston: Houghton Mifflin Co., 1948.
9. Stieglitz, Alfred. Quoted in *Masters of Photography* by Beaumont Newhall. New York: Castle Books, 1958, p. 172.
10. *Modern Photography* (December 1953): 54.
11. Hall, James Baker. *Rites and Passages*. Millerton: Aperture, 1978, p. 18.
12. Ibid.
13. Ibid.
14. Capo, B. *Paul Caponigro*. Millerton: Aperture, 1972, p. 42.
15. Emerson, Peter Henry. From Nancy Newhall. *P.H. Emerson—The Fight for Photography as a Fine Art*, Millerton: Aperture, 1975, p. 259.

6 Portraiture

1. Newhall, Beaumont. "First American Masters of the Camera," *Art News Annual* (1948): 91-98, 168-172.
2. Southworth, Albert Sands. *The British Journal of Photography* 18 (1871): 530-532.
3. Baudelaire, Charles. *The Mirror of Art*. New York: Phaidon Press Ltd., 1955, 228-231.
4. Cameron, Julia Margaret. "Annals of My Glass House." *Photo Beacon* (1890): 157-160.
5. Gernsheim, Helmut. *Julia Margaret Cameron*. New York: Fountain Press, 1948.
6. Cameron, Julia Margaret. "Annals of My Glass House." *Photo Beacon* (1890): 157-160.
7. Newhall, Nancy. "Emerson's Bombshell." *Photography* 1 (1947): 114.
8. Kramer, Robert. *August Sander: Photographs of an Epoch*. Millerton: Aperture, 1980, p. 27.
9. Mrázková, Daniela. *Masters of Photography*. New York: Exeter Press, 1987, p. 158.
10. Sherman, Cindy. *Cindy Sherman*. New York: Pantheon Books, 1984, back cover.
11. Donkins, Lydia Louise Summerhouse. Cited in "Annals of My Glass House," *Photo Beacon* (1890): 157-160.

7 Photojournalism

1. Hine, Lewis. Cited by Susan Sontag. *On Photography*, New York: Delta Press, 1977, p. 185.
2. Linton, W. J. *Wood Engraving, a Manual of Instruction*. London: George Bell and Sons, 1884.
3. *Scribner's Monthly* (March 1880) Editorial.
4. *New York Sun*, 12 February 1888.
5. *Early Photojournalism in Europe and America*. New York: Media Loft, 1977.
6. Phillips, Sandra. "The French Magazine *Vu*." *Picture Magazines Before "Life."* Woodstock, New York: 1982.
7. Luce, Henry. Cited by Arthur Rothstein. *Documentary Photography*. Boston: Focal Press, 1986, p. 79.
8. Luce, Henry. *Life* Prospectus, 1934.
9. Editors. "Introduction." *Life* 1 (1936).
10. Szarkowski, John. *Looking at Photographs*. New York: Museum of Modern Art, 1973, p. 154.
11. Smith, W. Eugene. Interview by Paul Hill and Thomas Cooper. *Dialogue with Photography*. New York: Farrar, Straus, and Giroux, 1979, p. 258.
12. Ibid., pp. 258-259.
13. Luce, Henry. *Life* Prospectus, 1934.
14. Edom, Clifton. *Photojournalism, Principles and Practice*. Dubuque: W.C. Brown, 1980, p. 107.
15. Gallagher, Wesley. Associated Press Brochure, 1973.
16. Chapnick, Howard. Quoted by Marianne Fulton. *Eyes of Time*. New York: New York Graphic Society, 1990, Foreword.

8 Alfred Stieglitz and the Photo-Secession

1. Kerfoot, J.B. *Camera Work* 8 (1904).
2. Newhall, Beaumont and Nancy. *Masters of Photography*. New York: Castle Books, 1958, p. 60.
3. Stieglitz, Alfred. *American Annual Photographer* (1895): 27-28.
4. Stieglitz, Alfred. *Camera Notes*. 6 (1902).
5. *Camera Notes*. 2 (July 1898).
6. Keiley, Joseph T. "The Photo-Secession Exhibition at the Pennsylvania Academy of Fine Arts—Its Place and Significance in the Progress of Pictorial Photography." In *Camera Work: A Critical Anthology*, edited by Jonathon Green. Millerton: Aperture, 1973, p. 102.
7. Stieglitz, Alfred. *Twice a Year* No. 8-9 (1942): 117.
8. *Camera Work: A Critical Anthology*, edited by Jonathon Green. Millerton: Aperture, 1973, p. 103.
9. Stieglitz, Alfred. *Camera Work* 1 (1903).
10. Kerfoot, J.B. *Camera Work* 8 (1904).
11. Stieglitz, Alfred. *The Photo-Secession* 1 (1902).
12. Stieglitz, Alfred. *Camera Work* 18 (1907).

13. Chamberlain, J.E. "*N.Y. Evening Mail.*" *Camera Work* 23 (1908).
14. Ibid.
15. Stieglitz, Alfred. *Photo Miniature* 124 (March 1913).
16. Stieglitz, Alfred. *Camera Work* 48 (1916).
17. Ibid.
18. Newhall, Beaumont and Nancy. *Masters of Photography*. New York: Castle Books, 1958, p. 60.
19. Ibid.
20. Stieglitz, Alfred. *Twice a Year* (1943): 252.

9 Farm Security Administration (FSA)

1. Stryker, Roy and Wood, Nancy. *In This Proud Land*. New York: New York Graphic Society, 1973, pp. 8-9.
2. Ibid., p. 11.
3. Ibid.
4. Ibid.
5. Ibid., p. 14.
6. Ibid.
7. Ibid.
8. Rosenblum, Naomi. *A World History of Photography*. New York: Abbeville Press, 1984, p. 383.
9. *Camera Arts* (February 1983): 34.
10. Ibid., p. 37.
11. Stryker, Roy and Wood, Nancy. *In This Proud Land*. New York: New York Graphic Society, 1973, p. 19.
12. Lorenz, Paré. "Dorthea Lange, Camera with a Purpose." *U.S. Camera* (1941): 93-98.
13. *Camera Arts* (February 1983): 83.
14. Stryker, Roy and Wood, Nancy. *In This Proud Land*. New York: New York Graphic Society, 1973, p. 17.
15. Ibid.

10 Form and Design

1. Metzker, Ray K. Conversation with Diana Johnson, December 6, 1977. Cited in *Spaces*, Rhode Island School of Design Museum Notes, 1978, p. 78.
2. Coburn, Alvin Langdon. Letter to Beaumont Newhall, January 15, 1963. Cited by Beaumont Newhall. *The History of Photography*. New York: Museum of Modern Art, 1982, p. 201.
3. *Pröun und Wolkenbugel*. Dresden: 1977, p. 65.
4. Weiss, Evelyn ed. *Alexander Rodchenko: Fotografien 1920–1938*. Dresden: Wienand Verlag, 1978, pp. 50-57.
5. Cartier-Bresson, Henri. *The Decisive Moment*. New York: Verve and Simon and Schuster, 1952.
6. Malcolm, Janet. "Two Roads, One Destination." In *Diana and Nikon*. Boston: David R. Godine, 1980, p. 114.
7. Moholy-Nagy, Laszlo. *Painting, Photography, Film*. Cambridge: MIT Press, 1973, p. 93.

11 Photography and Combat

1. Editorial. *New York Times*, 20 October 1862.
2. *American Journal of Photography* 5 (1862): 145.
3. Meredith, R. *Mr. Lincoln's Cameraman*. New York: Scribner's Press, 1946.
4. Hare, Jimmy. *News Photographer*. New York: MacMillan Press, 1940, p. 15.
5. Mott, Frank. *American Journalism: A History of Newspapers in the United States Through 250 Years 1690–1940*. New York: MacMillan Press, 1956, p. 529.
6. Lemagny, Jean Claude. *A History of Photography*. Cambridge: Cambridge University Press, 1986, p. 126.
7. Ibid.
8. Flora, Francesco. *Ritratto do un Ventennio*. Milan: Alfa, 1965, pp. 115-182.
9. Rothstein, Arthur. *Documentary Photography*. Boston: Focal Press, 1986, p. 95.
10. *Travels in Hyper Reality; Essays*. Translated by William Weaver. New York: Harcourt Brace Jovanovich, 1986, p. 216.
11. Rothstein, Arthur. *Documentary Photography*. Boston: Focal Press, 1986, p. 102.
12. Ibid., p. 93.
13. White, T. H. *The Book of Merlyn*. Austin: The University of Texas Press, 1988.

12 Manipulation

1. Robinson, H.P. *Pictorial Effect in Photography*. London: Piper and Carter, 1869, pp. 78 and 109.
2. Baudelaire, Charles. *Photography, the Mirror of Art*, London: Phaidon Press Ltd., 1955, pp. 228-231.
3. Rejlander, Gustave. Quoted in *The Practical Photographer* 6 (1895): 70.
4. *Literary Gazette* quoted in *The Practical Photographer* 6 (1895): 68-69.
5. *Webster's Seventh New Collegiate Dictionary*, S.V. "surrealism."
6. Foster, Hal. "Subversive Signs." *Art in America* (November 1982): 88.
7. Michaels, Duane. Quoted by Anne Hoy. *Fabrications*. New York: Abbeville Press, 1987, p. 191.

The amount of available information regarding photographic history is vast. While the author used numerous sources in writing this book, the bibliography lists only those volumes that have been consulted repeatedly.

Adams, Ansel. *Ansel Adams: An Autobiography*. New York: New York Graphic Society, 1985.
Beaton, Cecil, and Gail Buckland. *The Magic Image: The Genius of Photography from 1839 to the Present Day*. Boston: Little Brown and Company, 1975.
Buckland, Gail. *Fox Talbot and the Invention of Photography*. Boston: David R. Godine, 1980.
Coke, Van Deren. *The Painter and the Photograph from Delacroix to Warhol*. Revised Edition. Albuquerque: University of New Mexico Press, 1972.
Edom, Clifton. *Photojournalism: Principles and Practices*. Second Edition. Dubuque, Iowa: Wm. C. Brown Company, 1980.
Flukinger, Roy. *The Formative Decades, Photography in Great Britain 1839–1920*. Austin: University of Texas Press, 1985.
Fulton, Marianne. *Eyes of Time, Photojournalism in America*. Boston: Little Brown and Company, 1988.
Gernsheim, Helmut, and Alison Gernsheim. *The History of Photography from the Camera Obscura to the Beginning of the Modern Era*. New York: McGraw–Hill Book Co., 1969.
Gilbert, George. *Photography: The Early Years*. New York: Harper and Rowe, 1980.
Green, Jonathan. *Camera Work: A Critical Anthology*. New York: Aperture, 1973.
Hill, Paul and Thomas Cooper. *Dialogue with Photography*. New York: Farrar, Straus, Giroux, 1979.
Hoy, Anne. *Fabrications, Staged, Altered, and Appropriated Photographs*. New York: Abbeville Press, 1987.
Jussim, Estelle and Elizabeth Lindquist–Cock. *Landscape as Photograph*. New Haven: Yale University Press, 1985.
Kobre, Kenneth. *Photojournalism: The Professionals' Approach*. Stoneham, Mass.: Focal Press, 1980.
Leavens, Ileana. *From "291" to Zurich: The Birth of Dada*. Ann Arbor, Mich.: UMI Research Press, 1983.
Lemagny, Jean–Claude, and Andre Rouille. *A History of Photography*. Cambridge: Cambridge University Press, 1986.
Life Library of Photography. 17 Volumes. New York: Time–Life Books, 1970–1972.
Lyons, Nathan, ed. *Photographers on Photography: A Critical Anthology*. Englewood Cliffs, N.J.: Prentice–Hall, 1966.
Mrázková, Daniela. *Masters of Photography*. New York: Exeter Books, 1987.
Newhall, Beaumont. *The History of Photography*. New York: The Museum of Modern Art, 1982.
Newhall, Beaumont, ed. *Photography: Images and Essays*. New York: Museum of Modern Art, 1980.
Newhall, Beaumont and Nancy Newhall. *Masters of Photography*. New York: Castle Books, 1958.
Newhall, Nancy. *P.H. Emerson—The Fight for Photography as a Fine Art*. New York: Aperture, 1975.
Newhall, Nancy, ed. *Edward Weston: The Flame of Recognition. An Aperture Monograph*. New York: Grossman Publishers, 1971.
Rosenblum, Naomi. *A World History of Photography*. New York: Abbeville Press, 1984.
Rothstein, Arthur. *Documentary Photography*. Boston: Focal Press, 1986.
Stange, Maren. *Symbols of Ideal Life: Social Documentary Photography in America*. New York: Cambridge University Press, 1989.
Stryker, Roy, and Nancy Wood. *In This Proud Land—America 1935–1943 as Seen in the FSA Photographs*. Boston: New York Graphic Society, 1973.
Szarkowski, John. *Looking at Photographs*. New York, Museum of Modern Art, 1973.
Szarkowski, John. *Mirrors and Windows: American Photography Since 1960*. New York: Museum of Modern Art, 1978.
Taft, Robert. *Photography and the American Social Scene: A Social History 1839–1889*. New York: Dover Publications, 1964.
Weaver, Mike. *The Art of Photography*. New Haven: Yale University Press, 1989.

Key:

CCP
Center for Creative Photography
Tucson, Arizona

ICP
International Center of Photography
New York

IMP/GEH
International Museum of Photography
at the George Eastman House
Rochester, New York

MOMA
Museum of Modern Art
New York, New York

RISD
Rhode Island School of Design
Providence, Rhode Island

The following is a more complete guide to the credits that appear in figure legends. The alphabetized list includes individuals and institutions that provided photographs and/or permission to reproduce photographs from their respective collections. The location of images within the book is denoted through a system of two numbers. The first number refers to the appropriate chapter; the second number refers to the individual image's placement relative to other illustrations within the chapter. For example, 3–4 denotes that the image may be found as the fourth illustration within Chapter 3, "Camera and Canvas— The Alliance Between Painters and Photographers."

Sources of Imagery:

Associated Press/Wide World Photos, New York, New York: 11-1, 11-13.
Jayne Baum Gallery, New York: 6-18.
Bibliothéque Nationale, Paris: 6-5.
BLACK STAR, New York, New York: I-5.
David Winton Bell Gallery, Brown University, Providence, Rhode Island:
    Gift of Mr. and Mrs. Elmer Rigelhaupt: 3-4.
    Vera and Albert List Purchase Fund: 12-13.
Center for Creative Photography, Tucson, Arizona: 5-9, 5-10, Color Plate 8.
Chicago Historical Society: 2-5.
Imogen Cunningham Trust, Berkeley, California: 10-1.
The Fox Talbot Museum: 1-9.
Gernsheim Collection, Harry Ransom Humanities Research Center, The University of Texas at Austin: 1-1.
International Center of Photography, New York, New York: 4-12.
International Museum of Photography at the George Eastman House, Rochester, New York:
    I-1, 2-1, 2-3, 2-6, 3-1, 3-3, 4-8, 5-2, 5-13, 7-4, 8-2, 8-7, 10-3, 11-7, 12-6, Color Plate 2.
Library Company of Philadelphia: 6-6.
Library of Congress, Washington, D.C.: 9-1, 9-2, 9-3, 9-4, 9-5, 9-6, 9-7, 9-9.
MAGNUM Photos, New York, New York: 7-10, 10-6, 11-11, 11-12, 11-14.
Mary Ellen Mark/Library, New York, New York: 7-9.
Metro Pictures, New York, New York: 6-17.
Robert Miller Gallery, New York, New York: 4-15.
Museum of the City of New York: 7-2.
Museum of Fine Arts, Boston: 6-1.

Museum of Fine Arts, Houston, Texas: I-3.
Museum of Modern Art, New York, New York: 3-5, 4-4, 5-3, 8-5, 8-6, 11-6, 11-8, 12-8, 12-9, Color Plate 5.
Gift of the George Eastman House: 6-4.
Gift of Georgia O'Keeffe: 8-3.
Mr. and Mrs. John Spencer Fund: 10-2.
Gift of Shirley Burden: 6-9, and the Estate of Vera Louise Fraser: 11-5.
Gift of Edward Steichen: 6-10, 11-9.
Gift of Philadelphia Commercial Museum: 3-2.
Gift of Richard Avedon: 6-11.
Gift of Bruce Davidson: 6-13.
National Archives, Washington D.C.: 9-10.
New York Public Library/Astor, Lenox and Tilden Foundations
Photography Collection, Miriam and Ira D. Wallach Division of Art Prints and Photographs: 6-8.
General Research Division: 4-2.
The Oakland Museum, (Dorthea Lange Collection), Oakland, California: 9-8.
Pace/MacGill Gallery, New York: 3-6, 3-7.
Photo Researchers, New York, New York: 4-9.
The Art Museum, Minor White Archive, Princeton University, Princeton, New Jersey: 5-11.
Private Collections: 4-6, 6-14, 7-5, 10-4, 12-4, 12-10, 12-11, Color Plate 3, Color Plate 4, Color Plate 7.
Rhode Island School of Design, Museum of Art, Providence, Rhode Island
All photography by Cathy Carver: 1-10, 2-2, 6-7, 10-5, 12-3.
Purchased with the aid of funds from the National Endowment for the Arts:
4-10, 4-14, 12-1, 12-14.
Edgar J. Townes and Walter H. Kimball Funds: 4-1, 6-12, 8-4.
Gift of the Photography Department: 12-12, 12-15.
Gift of Roy Neuberger: 4-3.
Jesse Metcalf Fund: 4-5, 5-4, 5-6, 8-1.
Helen M. Danforth Fund: 5-7.
Gift of Mrs. Barnet Fain: 5-8, 7-1.
Gift of Mr. and Mrs. Leslie Seekell Fletcher: Color Plate 6.
Gift of Matrix Publications and Alpha Partners: 4-11.
Gift of Mr. and Mrs. Gilman Angier: 5-1, 6-15, In Memory of Julia Angier Ewing and
Colby MacKinney Keeler: 5-5.
Gift of Frederick J. Myerson: 4-13.
Gift of Aaron Siskind: 4-7.
Royal Photographic Society, Bath England: 12-5.
Science Museum, London: 11-4, (#124/64), Color Plate 1 (#30).
Société Française de Photographie, Paris: 1-8, 1-11.
Holly Solomon Gallery, New York, New York: 6-16.
Time/Life—Life Picture Service: I-4, 7-6, 7-7, 7-8, 11-10.
University of California at Los Angeles, Special Collections: 12-7.
University of Rochester: I-2.
Visual Studies Workshop Gallery: 5-12.